Physics and Chemistry in Space
Volume 7

Edited by
J. G. Roederer, Denver

M. Schulz · L. J. Lanzerotti

Particle Diffusion in the Radiation Belts

With 83 Figures

Springer-Verlag
New York Heidelberg Berlin 1974

Michael Schulz

The Aerospace Corporation, Los Angeles, CA 90009/USA

Louis J. Lanzerotti

Bell Telephone Laboratories, Murray Hill, NJ 07974/USA

The illustration on the cover is adapted from Figure 58, which shows the evolution of an inner-zone electron-flux profile ($E > 1.9$ MeV, omnidirectional) observed on Explorer 15 following the high-altitude nuclear explosion of 1 November 1962. The data demonstrate the simultaneous effects of radial diffusion and pitch-angle diffusion.

ISBN 0–387–06398–6 Springer-Verlag New York Heidelberg Berlin
ISBN 3–540–06398–6 Springer-Verlag Berlin Heidelberg New York

Monophoto typesetting and offset printing: Zechnersche Buchdruckerei, Speyer.
Bookbinding: Konrad Triltsch, Würzburg.

Preface

The advent of artificial earth satellites in 1957–58 opened a new dimension in the field of geophysical exploration. Discovery of the earth's radiation belts, consisting of energetic electrons and ions (chiefly protons) trapped by the geomagnetic field, followed almost immediately [1, 2]. This largely unexpected development spurred a continuing interest in magnetospheric exploration, which so far has led to the launching of several hundred carefully instrumented spacecraft.

Since their discovery, the radiation belts have been a subject of intensive theoretical analysis also. Over the years, a semiquantitative understanding of the governing dynamical processes has gradually evolved. The underlying kinematical framework of radiation-belt theory is given by the adiabatic theory of charged-particle motion [3], and the interesting dynamical phenomena are associated with the violation of one or more of the kinematical invariants of adiabatic motion.

Among the most important of the operative dynamical processes are those that act in a stochastic manner upon the radiation-belt particles. Such stochastic processes lead to the diffusion of particle distributions with respect to the adiabatic invariants. The observational data indicate that some form of particle diffusion plays an essential role in virtually every aspect of the radiation belts.

With radiation-belt physics now in its second decade, it seems desirable that the existing observational and theoretical knowledge of radiation-belt dynamics be consolidated into a unified presentation. This indeed is one purpose of the present series of monographs. Two of the earlier volumes of *Physics and Chemistry in Space* have treated the subjects of geomagnetic micropulsations [4] and adiabatic charged-particle motion in the magnetosphere [5], respectively. The present volume is concerned principally with the diffusion processes that affect radiation-belt particles. Some of these diffusion processes can cause energetic particles to precipitate into the atmosphere, thereby creating the aurora, as described in the fourth volume of this series [6].

Readers may find previous volumes of this series useful references for a more elaborate discussion of geomagnetic-activity indices [4] and adiabatic motion [5] than can be given here. To the extent that adiabatic invariants are discussed in the present volume (Chapter I), they are treated from the Hamiltonian viewpoint [7], since this formulation is especially suitable for subsequent application to the analysis of particle diffusion. In this and other respects, an effort has been made to present the existing state of knowledge in a manner that is instructively novel.

Radiation-belt physics is especially rewarding in that it requires a close collaboration of ideas between the observational and theoretical spheres. Perhaps the necessity for this collaboration arises from the fact that (as in any environmental science) there can be no hope of controlling the entire magnetosphere by experimental technique, or even of measuring *all* the parameters that may have a bearing on the dynamical phenomena. By the same token, the environment is suffi-ciently complex that there can be no hope of theoretically predicting radiation-belt behavior solely from a set of mathematical postulates. Instead, the theoretical groundwork only makes it possible to relate one set of observations to another. The governing diffusion equations may be known in analytical form, but the transport coefficients that enter them are essentially empirical.

The principal purposes of this book are (1) to convey a quantitative understanding of the fundamental ideas prevalent in radiation-belt theo-ry, and (2) to instruct the reader in how to recognize the various diffusion processes in observational data, and how then to extract numerical values for the relevant transport coefficients. Since the subject is rather specialized, the book will be of value primarily to those who have a professional interest in the subject. This prime audience would include active researchers in the field of space physics, post-doctoral scientists redirecting their interests from another branch of physics, and graduate students seeking to do thesis research on the topic of radiation-belt dynamics. The book is probably unsuitable as a graduate-course textbook (unless used in conjunction with other volumes in the series), but should serve well as a reference work. Moreover, although the applications specifically cited refer to the earth's magnetosphere, many of the basic ideas are current in the understanding of laboratory-plasma devices and non-terrestrial magnetospheres.

While the present volume is essentially self-contained, some sections offer much more difficult reading than others. The reader should feel free to pass over those portions that least interest him, accepting on faith (where necessary) the validity of such intermediate results that find subsequent application. For the more critical reader, an attempt has been made to indicate the basic justification for the various analytical procedures, although little attention has been given to mathematical rigor in the formal sense.

References to the literature are numbered approximately in order of first citation. These numbers (in square brackets) are keyed to a list that appears at the end of the volume. The length of the reference list has been held to the minimum consistent with the avoidance of plagiarism. The list includes a few truly classic papers in the field, a few sources of supplementary information too complicated for inclusion here, and the sources of observational and theoretical data used in figures and tables, or otherwise invoked in support of an argument. The list is therefore not representative of the excellent contributions of many investigators, and is not intended to be used for scorekeeping purposes.

The authors are pleased to thank their many colleagues whose suggestions have helped to optimize the presentation of current ideas in this monograph, especially Dr. J. G. Roederer, Dr. J. M. Cornwall, Dr. T. A. Farley, Dr. M. Walt, Dr. G. A. Paulikas, Dr. J. B. Blake, Dr. W. L. Brown, Dr. H. C. Koons, Dr. A. Eviatar, Dr. Y. T. Chiu, Dr. A. L. Vampola, Dr. D. P. Stern, Dr. A. Hasegawa, and Ms. C. G. Maclennan. Finally, it is a pleasure to thank Miss Doreen Bracht for typing the final manuscript.

August 1973   Michael Schulz

                                                            L. J. Lanzerotti

Contents

Contents

Introduction

The earth's radiation belts consist of energetic electrons and ions (mostly protons) whose motion is controlled by the geomagnetic field. Radiation-belt particles typically have energies ranging upward from 100 keV, and each is constrained by the field to execute a rather complicated orbital motion that encircles the earth with a characteristic *drift period* (typically tens of minutes). Thus, the particles are magnetically trapped by the earth's field.

Ideally, the radiation belts are mapped by means of satellite-borne sensors that distinguish between particle species and count the differential unidirectional particle flux at each energy and pitch angle of interest, and at all points in space. Practical considerations generally limit the available information quite severely, but a reasonable picture of the earth's radiation environment can be assembled from the available data. For example, Fig. 1 provides a contour plot of the observed *omnidirectional* electron flux (integrated over all directions of incidence relative

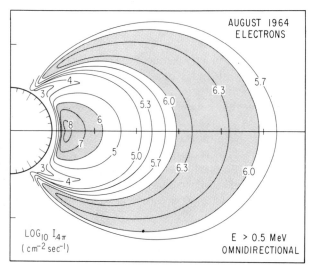

Fig. 1. Contours of constant integral electron flux $I_{4\pi}$ in the earth's radiation environment [8]. Shaded regions correspond to $I_{4\pi} > 10^6$ cm^{-2} sec^{-1}.

to the local magnetic field) at energies above 0.5 MeV [8]. This coverage extends over a considerable region of space and clearly defines a pair of toroids (known as the *inner belt* and the *outer belt*) in which the radiation intensity concentrates. The origin of the so-called *slot* between the two belts need not be of immediate concern here, as the same dynamical principles govern both the inner belt and the outer.

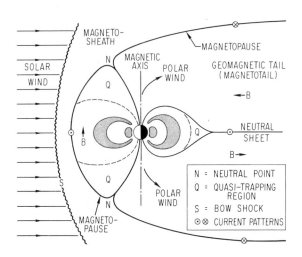

Fig. 2. Magnetospheric configuration in the noon-midnight meridional plane. Shaded regions correspond to the radiation belts of Fig. 1. Dashed curves represent boundaries of quasi-trapping regions.

The trapped-radiation environment is not observed to extend to field lines that attain altitudes greater than about seven earth radii (equatorial geocentric distances of about eight earth radii). The underlying explanation for this cut-off is that the *solar wind*[1] confines the geomagnetic field to a cavity known as the *magnetosphere*. Figure 2 illustrates some of the important features of the magnetosphere. The boundary of this cavity is typically located at an equatorial geocentric distance ~ 10 earth radii on the noon meridian. At equatorial geocentric distances beyond about eight earth radii, the magnetic field is so distorted that it typically cannot support a completely closed particle drift orbit.

---

[1]The solar wind is a supersonic plasma flow of solar-coronal origin that typically attains a velocity ~ 300 km/sec and a density ~ 8 electrons/cm^3 at heliocentric distances ~ 1 AU [9]. The ionic constituents are chiefly protons, with a small ($\sim 4\%$) admixture of alpha particles. All these parameters vary considerably with time, but in a manner consistent with gross charge quasi-neutrality.

Field lines that pass within about two earth radii of the boundary thus comprise what is substantially a *quasi-trapping zone*, in the sense that particles injected there typically cannot complete an azimuthal drift orbit of 2π radians before escaping across the magnetospheric surface to interplanetary space. More precisely, there exist so-called *quasi-trapping regions* on such field lines; particles traveling perpendicular to the local field in these regions are incapable of executing a complete drift orbit [5].

The magnetospheric surface consists of the magnetopause and the neutral sheet (see Fig. 2). The *magnetopause* is that portion of the magnetospheric surface upon which the solar wind impinges after passing through the *bow shock*[2]. At points immediately interior to the magnetopause, the geomagnetic field has an intensity just sufficient to balance the dynamical pressure imparted by the incident solar wind (other contributions to this pressure balance being negligible). The magnetopause includes a pair of *neutral points* at high latitudes on the noon meridian. These points, at which the field ideally vanishes because the shocked solar wind passes by undeflected, separate the field lines that emanate from the earth's surface on the noon meridian into two classes, *viz.*, field lines that close through the equatorial plane on the day side, and those that extend into the *geomagnetic tail*. The tail extends to large distances (at least hundreds of earth radii) in the antisolar direction, beyond which the field lines either close through the equatorial plane (closed model of the magnetosphere) or merge with lines of the interplanetary field (open model). Since the field lines in the two halves of the tail originate from opposite polar caps of the earth, there must be a region of field reversal in between. This region is idealized as a *neutral sheet*, which extends approximately in a plane across the entire magnetotail and joins with the magnetopause at the eastern and western flanks. The geomagnetic tail is maintained electrodynamically by a current that flows westward along the neutral sheet and returns eastward over the northern and southern flanks of the magnetopause.

Since the magnetic field in the immediate vicinity of the neutral sheet is very weak (even when compared with the interplanetary field $\sim 5\gamma$, where $1\gamma \equiv 10^{-5}$ gauss), plasma tends to concentrate in this region that separates the two halves of the magnetotail. Thus, the magnetic

---

[2]Since the magnetosphere plays the role of a blunt object inserted in a supersonic flow (the solar wind), the shock is detached (separated) from the magnetosphere. The separation amounts to ~ 2 earth radii along the earth-sun axis [10]. The region between the bow shock and the magnetopause is known as the *magnetosheath* and is characterized as a region of considerable plasma turbulence.

neutral sheet is immersed in a *plasma sheet*, whose currents are in fact those responsible for maintaining the magnetic configuration described above. Various collective excitations of this plasma lead to a *nonvanishing* electrical resistivity, and so an electrostatic field appears across the nightside magnetosphere from east to west. In reality, this field extends across the dayside magnetosphere as well, and leads to a sunward convection of magnetospheric plasma [11] by virtue of the $E \times B$ drift (see below). This electrostatic field is therefore called the *convection electric field*; it extends across the magnetosphere from dawn to dusk. The sunward plasma flow caused by this electric field counterbalances an outward flow induced in a "viscous" boundary layer (~ 100 km thick) by a plasma-kinetic interaction between the solar wind and the magnetospheric plasma [12]. It is this "viscous" interaction (which probably involves several distinct plasma wave modes and instabilities) that is ultimately responsible (quite indirectly) for formation of the neutral sheet and geomagnetic tail.

As suggested above, the plasma sheet may be a repository for plasma of interplanetary origin. In addition, hydrogen plasma from each polar ionosphere is believed to flow outward along field lines that extend into the magnetotail. This flow, known as the *polar wind*, results from an ambipolar electric field established by electrons, protons, and oxygen ions (O^+) in the ionosphere. The ambipolar field is sufficiently strong to eject the lighter ions (H^+) against the restraint of gravity. It is presumed that some portion of the polar wind enters the plasma sheet, while the remainder escapes to interplanetary space.

Were it not for the rotation of the earth, all cold magnetospheric plasma of ionospheric origin would also escape. The sunward flow induced by the convection electric field would transport this plasma either directly to interplanetary space, or at least to the "viscous" boundary layer of the magnetosphere (see above). In fact, the earth's rotation creates an electrostatic field whose equipotential surfaces form closed shells. Cold plasma (by definition) follows trajectories that coincide with equipotential surfaces of the superimposed electric fields. Some of these surfaces form closed shells only weakly distorted from equipotentials of the *corotation electric field*. Others form open surfaces only weakly distorted from equipotentials of the convection electric field, and these intersect the magnetopause. The open equipotential surfaces intersect the ionosphere at high latitudes, but below the limit of open (to the tail) field lines that define the magnetic polar cap and carry the polar wind. Ionospheric plasma from these latitudes flows upward into the magnetosphere as the polar wind does, but is removed from the magnetosphere by convection across closed field lines to the boundary. The result is a high-latitude *trough* in ionospheric density, and

the associated flow is sometimes called the *trough wind*. The upward
flow velocity here is generally smaller than in the polar wind.

Closed equipotential surfaces of the magnetospheric electric field
(convection plus corotation) intersect the ionosphere at low and middle
latitudes. Since cold plasma from the ionosphere cannot escape the
magnetosphere either along open field lines or open equipotentials from
this region, the associated magnetospheric plasma can build up an
appreciable density and attain diffusive equilibrium with the ionosphere.
In fact, the "last" closed equipotential surface is closely associated with
a morphological feature known as the *plasmapause*, across which the
observed plasma density can vary by three orders of magnitude over
a distance ~ 1000 km in the equatorial plane. Beyond the plasmapause
a cold-plasma density ~ 1 cm^{-3} is typical. Within the *plasmasphere*,
which is bounded by the plasmapause, densities $\sim 10^3$ to 10^4 cm^{-3}

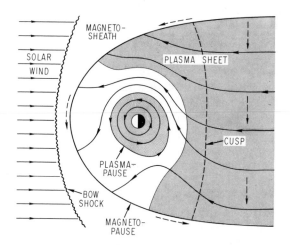

Fig. 3. Distribution (shaded areas) and flow pattern (solid arrows) of magnetospheric
plasma ($E \lesssim 10$ keV) in the equatorial plane. Dashed arrows indicate current pattern.
The plasmasphere is the shaded region inside the plasmapause.

are usual [13]. Figure 3 is a polar view of the magnetospheric equatorial
plane in which the various electric equipotentials and regions of plasma
concentration are indicated schematically[3].

---

[3]While the plasmasphere contains truly cold plasma (temperature ~ 1 eV),
the plasma sheet has a proton temperature ~ 6 keV.

As distinguished from cold plasma, there exists a class of particles (hot plasma) whose motion is governed not only by magnetospheric convection and corotation fields, but also by drifts related to the gradient and curvature of the local magnetic field **B**. By contrast with the energy- and charge-independent drift caused by an electric field **E**, plasma drifts related to $\nabla_\perp B$ and $\partial\hat{\mathbf{B}}/\partial s$ (gradient and curvature, where s is a coordinate measured along **B**) increase monotonically with particle energy and proceed in a direction determined by the signature of the particle charge q. As a rather general rule, application of a force **F** to a particle confined by a magnetic field results in a drift velocity $\mathbf{v}_d = (c/q B^2)\mathbf{F}\times\mathbf{B}$, where c is the speed of light (see Section III.6). In case $\mathbf{F}=q\mathbf{E}$, this formula is valid whenever it predicts $v_d < c$. Otherwise, the requirement is that $v_d \ll v$, where **v** is the total velocity of the particle. Curvature drift, for example, is driven by the centrifugal force $\mathbf{F}=mv_\parallel^2\,(\partial\hat{\mathbf{B}}/\partial s)$, where m is the particle mass and $v_\parallel = \mathbf{v}\cdot\hat{\mathbf{B}}$.

A typical magnitude for the convection electric field is $5\,\mu\text{V/cm}$ [14]. On an azimuthal-drift orbit having a diameter ~ 8 earth radii ($\sim 5\times10^9$ cm), therefore, a typical particle energy ~ 25 keV might characterize the hot plasma, for which electric and gradient-curvature drifts are supposed to be of comparable importance. In fact, the majority of magnetospheric particle energy resides in protons of this type, the spatial distribution of which often exhibits a peak density $\gtrsim 1\,\text{cm}^{-3}$. The gyration of these particles in a field $\sim 100\,\gamma$ can produce a significant diamagnetic effect locally, and the *ring current* associated with their drifts (and modified by diamagnetic effects) can produce a field depression observable even at the earth's surface [15]. Electrons ~ 25 keV in energy are found to produce a smaller, but still significant, contribution to the ring current (perhaps 25% of the total). Trajectories of ring-current particles (Fig. 4) can follow either open or closed surfaces, depending upon energy, particle species, and location in space [5]. This is true even though the field lines on which ring-current (hot-plasma) particles reside typically generate closed drift shells for particles of much larger energy (*e. g.*, > 1 MeV).

In practice, the hot magnetospheric plasma is easily distinguished from the coexisting cold plasma, which has a temperature ~ 1 eV [16]. In view of the typical electric-field magnitude ($5\,\mu\text{V/cm}$), gradient and curvature drifts must be negligible for particle energies $\lesssim 1$ keV. The demarcation between ring-current (hot-plasma) and radiation-belt particles is somewhat more nebulous. Ideally, a radiation-belt particle should be sufficiently energetic that the steady magnetospheric electric field exerts no significant influence on the drift trajectory. In this limit the drift shell, generated by specifying an initial location in space and local pitch angle $\cos^{-1}(\hat{\mathbf{v}}\cdot\hat{\mathbf{B}})$, is independent of particle energy. Moreover,

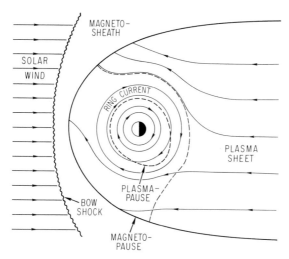

Fig. 4. Adiabatic flow pattern of magnetospheric protons ($E \sim 10-200$ keV) in the equatorial plane. Dashed curves represent boundaries of plasma sheet and plasmasphere.

the *mirror field* (denoted B_m), i. e., the magnitude of **B** at which $\hat{\mathbf{v}} \cdot \hat{\mathbf{B}} = 0$, remains constant as the particle drifts in azimuth. Thus, a radiation-belt particle for which **v** remains perpendicular to **B** at all times will experience only gradient drift, and so its drift motion will trace a contour of constant magnetic-field intensity on the equatorial surface (Fig. 5). From this viewpoint, the magnetic equator is the set of all points at which the magnitude of **B** attains a relative minimum with respect to displacement along the field line, i. e., the set of points for which $\partial B/\partial s = 0$ and $\partial^2 B/\partial s^2 > 0$.

Strictly speaking, the convenient idealization of ignoring the steady **E** field should be applicable only for particle energies $\gtrsim 200$ keV. It is customary, however, to analyze charged-particle behavior not only in this limit, but also for energies as low as ~ 40 keV, within the general framework of radiation-belt theory. For many purposes the steady magnetospheric electric field plays a complicating but nonessential role in the analysis at energies in the range $40-200$ keV. In other situations the inclusion of **E** is absolutely necessary. The degree of analytical sophistication required in the treatment of any given topic in radiation-belt dynamics is usually indicated by the quality of the available data. The question of whether to include **E** falls within this context.

The dynamical behavior of radiation-belt particles is most conveniently expressed and most easily visualized within the kinematical framework of the adiabatic theory of charged-particle motion [3]. A

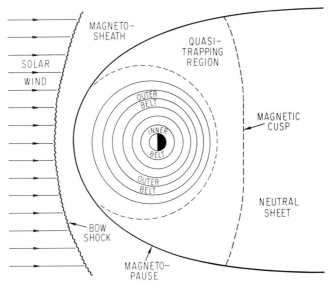

Fig. 5. Adiabatic drift paths of equatorially mirroring charged particles ($E \gtrsim 200\,\mathrm{keV}$). Protons drift westward (clockwise) and electrons eastward.

particle whose azimuthal drift motion defines a closed shell coincident with closed field lines at all longitudes can be assigned a set of three *adiabatic invariants*, which are (apart from dimensional constants) identifiable with the canonical momenta (action variables) of Hamilton-Jacobi theory [7]. As long as the particle remains undisturbed by forces varying suddenly on the drift time scale, the three adiabatic invariants are conserved quantities. The transformation to action-angle variables thus provides a kinematical framework in which the static and quasi-static characteristics of the magnetosphere are suppressed. Preservation of the adiabatic invariants is analogous to force-free rectilinear motion, in which the three components of linear momentum are separately conserved.

Geophysically interesting dynamical phenomena related to radiation-belt physics involve the violation of one or more adiabatic invariants. Such violation can be caused by particle-particle collisions, by wave-particle interactions, or by sufficiently sudden temporal changes in magnetospheric fields. Adiabatic invariants can be violated separately or in combination, depending upon the spatial structure and temporal character of the disturbing forces. Radiation-belt particles can be active conspirators in creating the disturbing forces, or they can be passive victims (test particles) to forces not of their own making. The various non-adiabatic perturbations that act upon geomagnetically trapped particles are unified, however, by an essential element of randomness, in that they distinguish among otherwise equivalent particles according to the instantaneous value (mod 2π) of a phase variable associated with the quasi-periodic unperturbed (adiabatic) motion.

Since the conditions of observation typically preclude the distinction of particle phases over time scales longer than a few drift periods, the day-to-day evolution of the earth's radiation belts cannot be traced with deterministic detail. Thus, for essentially all practical purposes, radiation-belt observations are phase-averaged. Moreover, the phase-averaged non-adiabatic evolution of the radiation belts proceeds via Brownian motion in canonical-momentum space, *i.e.*, by diffusion of the particle distribution among values of the adiabatic invariants.

Because particle diffusion in the radiation belts is best formulated in terms of the adiabatic invariants, a working knowledge of adiabatic theory is prerequisite to an understanding of particle diffusion. For this reason, the entire first chapter of the present volume is devoted to the adiabatic motion of charged particles, as applied to simple models of the earth's magnetosphere. The second and third chapters cover the theory of selected dynamical processes that violate adiabatic invariants and consequently lead to particle diffusion. The fourth and fifth chapters describe the interpretation and analysis of observational data having quantitative significance in terms of radial and pitch-angle diffusion, and the final (sixth) chapter briefly summarizes the present state of knowledge concerning particle diffusion in the radiation belts.

I. Adiabatic Invariants and Magnetospheric Models

I.1 Preliminary Considerations

As long as it remains trapped within the earth's magnetosphere, a radiation-belt particle of mass m and charge q typically executes a hierarchy of three distinct forms of quasi-periodic motion. The most rapid of these is *gyration* about a field line at frequency $\Omega_1/2\pi = -qB/2\pi mc$, where **B** is the local magnetic-field intensity and c is the speed of light. The instantaneous center of the gyration orbit is known as the particle's *guiding center*. An average over gyration reveals an oscillation of the guiding center between magnetic *mirror points*, which are located at a pair of well-defined positions along a path that very nearly coincides with the original field line. This periodic *bounce motion* between mirror points proceeds at a frequency $\Omega_2/2\pi \sim v/2\pi S$, where v is the speed of the particle and S is the arc length of the entire field line. Finally, an average over the bounce motion reveals an *azimuthal drift* of the guiding-center trajectory. This drift motion generates a shell encircling the earth; complete circuits of this shell are accomplished at the drift frequency $\Omega_3/2\pi \sim \langle v^2/2\pi\Omega_1 S^2 \rangle$, where the angle brackets denote an average over the entire particle orbit.

The time scales for gyration, bounce, and drift are respectively separated by a factor of order $\varepsilon \equiv \langle v/\Omega_1 S \rangle$. The limit $|\varepsilon| \ll 1$ is required for performing the averages mentioned above so as to separate the motion into the three distinct components. This condition imposes the requirement that the gyroradius be much smaller everywhere than the length of the guiding field line. Radiation-belt particles of interest here are thus distinguished by the requirement $|\varepsilon| \ll 1$ from very energetic particles such as galactic cosmic rays, which may have gyroradii as large as the magnetosphere. This limitation on radiation-belt energies is not a universally accepted convention, but it is conceptually useful to restrict the radiation belts to particles whose kinematics fall within the hierarchy outlined above. Special methods of numerical analysis [17] beyond the scope of the present treatment must generally be employed where $|\varepsilon| \gtrsim 1$. Such methods trace the details of each particle trajectory.

The limits on particle energy appropriate to the radiation belts can be estimated by calculating ε for a special class of particles, *viz.*, the class of particles magnetically confined to the equatorial plane of a dipole field. These particles can be thought of as "mirroring at the equator", or as bouncing with infinitesimal amplitude but finite frequency. If a is the radius of the earth (in which the magnetic dipole is assumed to be centered), the length of a field line that crosses the equatorial plane at a distance of L earth radii from the center is given (see Section I.4) by

$$S = 2La[1 + (1/2\sqrt{3})\ln(2 + \sqrt{3})] \approx 2.7603\,La. \qquad (1.01)$$

For a particle carrying the charge of Z protons, it follows that

$$\varepsilon \approx \beta L^2(mc^2/2.76\,qB_0a) \approx \beta L^2(mc^2/216\,Z\,\text{GeV}), \qquad (1.02)$$

where $\beta = v/c$, c is the speed of light, and $B_0 (\approx 0.31$ gauss$)$ is the equatorial magnetic-field intensity at the earth's surface. The limit $|\varepsilon| \ll 1$ required of radiation-belt particles is therefore satisfied by kinetic energies up to approximately $10/L^2\,\text{GeV}$ for protons, alpha particles, and other light ions, as well as relativistic electrons.

I.2 Action-Angle Variables

The three distinct periodicities associated with gyration, bounce, and drift motion give rise to a hierarchy consisting of three pairs of action-angle variables. The action variables J_i $(i = 1,2,3)$ are canonically defined [7] by the path integrals

$$J_i = \oint_i [\mathbf{p} + (q/c)\mathbf{A}] \cdot d\mathbf{l}, \qquad (1.03)$$

where $\mathbf{p}$ is the particle momentum and $\mathbf{A}$ is the electromagnetic vector potential. The first action integral J_1 corresponds to gyration about a field line. The first term of (1.03) for $i = 1$ is therefore equal to $2\pi p_\perp^2/m|\Omega_1|$, where $p_\perp$ is the component of $\mathbf{p}$ normal to $\mathbf{B}$. This follows from the fact that the orbit of gyration has a circumference equal to $2\pi v_\perp/|\Omega_1|$, where $v_\perp = p_\perp/m$. The second term of (1.03) for $i = 1$ is equal to q/c times the magnetic flux enclosed by the orbit of gyration. The net result is that

$$J_1 = \pi p_\perp^2 c/B|q| \qquad (1.04)$$

if one observes the sign convention that Ω_1 is positive for electrons, which gyrate in a positive (counterclockwise) sense about the field line. Since the rest mass m_0 is a constant of the motion, it is usual to extract from (1.04) a quantity

$$M \equiv p_\perp^2/2m_0 B, \qquad (1.05)$$

known as the *first adiabatic invariant*. This is not empty nomenclature, for M is indeed an invariant of the motion if the fields (*e.g.*, magnetic, electric, gravitational) seen by the particle remain virtually constant in time over the entire orbit of gyration. In practice this requires also that such fields do not vary significantly on a spatial scale as small as $v_\perp/|\Omega_1|$. Note that M/γ is equal to the magnetic moment of the particle, where γ ($= m/m_0$) is the usual relativistic factor.

The second action integral J_2 is evaluated along the bounce path, which is essentially parallel to the guiding field line and therefore encloses no magnetic flux. Convention does not distinguish between the second action integral J_2 and the *second adiabatic invariant*

$$J \equiv J_2 = \oint p_\| \, ds, \qquad (1.06)$$

where $p_\|$ is the component of $\mathbf{p}$ parallel to $\mathbf{B}$ and s is a curvilinear coordinate that measures distance along a field line from the equator. The adiabatic invariance of J holds for a particle acted upon by forces that remain virtually constant in time over the bounce period.

The third action integral J_3 is associated with the azimuthal drift motion. The integral around the drift shell may be evaluated along any closed curve that lies entirely on this surface and encircles the earth. For this action integral the first term of (1.03) is of order ε^2 (and therefore negligible) compared to the second. It follows that

$$J_3 = (q/c)\Phi, \qquad (1.07)$$

where Φ, the magnetic flux enclosed by the drift shell, is known as the *third adiabatic invariant*. The integral is independent of the path within the limitations specified above because no field lines intersect the drift shell in the limit $|\varepsilon| \ll 1$. The sign convention adopted in (1.07) corresponds to that for Ω_1, since the *drift* of electrons is also counterclockwise. Thus, the signature of Φ is positive for $q > 0$ and negative for $q < 0$. The third invariant is generally conserved for a particle acted upon by forces that remain virtually constant in time over the complete drift period. Figure 6 provides contour plots of the gyration, bounce, and drift frequencies versus kinetic energy and L for protons and electrons mirroring at the equator of a geomagnetic dipole field [see Section I.4].

By their execution of all three types of adiabatic motion, particles that belong to the radiation belts are distinguished from a variety of other particles found within the magnetosphere. Thus, solar cosmic-ray particles having energies appropriate to the radiation belts often enter the geomagnetic tail and descend to the polar caps. Since the tail does not support bounce motion, however, these particles must either precipitate into the polar atmosphere or mirror magnetically and return

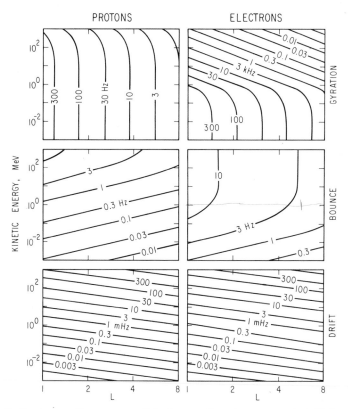

Fig. 6. Contours of constant adiabatic gyration, bounce, and drift frequency for equatorially mirroring particles in a dipole field. Adiabatic approximation fails in upper right-hand corners ($E \sim 1\,\mathrm{GeV}$, $L \sim 8$), since $\Omega_1 \sim \Omega_2 \sim \Omega_3$ implies $|\varepsilon| \sim 1$.

to interplanetary space. As they do not remain trapped within the magnetosphere, these particles disappear from the polar caps as soon as their immediate source (*e. g.*, a solar flare) is extinguished.

Particles that populate the quasi-trapping regions (see Introduction) are similarly excluded from the radiation belts by their inability to complete a drift period. A quasi-trapping region supports bounce motion and yields a well-defined second invariant, but it generates only partial drift shells that intersect the magnetospheric surface either at the magnetopause or at the neutral sheet. Thus, particles whose mirror points lie within a quasi-trapping region do not persist after withdrawal of their source and (by this convention) do not belong to the radiation belts.

For those particles that do execute all three types of quasi-periodic motion, the quantities $J_i/2\pi$ $(i=1,2,3)$ constitute a complete set of canonical angular momenta. The angular coordinates to which the $J_i/2\pi$ are conjugate can be identified as the phases φ_i that describe the progress made by a particle toward the completion of a gyration, bounce period, and drift period. Each phase φ_i is considered to advance at its own characteristic rate Ω_i, so as to achieve an increment of 2π upon completion of the period $2\pi/\Omega_i$ of the motion.

Conservation of the adiabatic invariants M, J, and Φ requires, in effect, that φ_1, φ_2, and φ_3 be cyclic coordinates [7] of the dynamical problem. Violation of the invariants occurs only in the presence of forces that vary on a sufficiently short spatial or temporal scale that particles having different phases respond differently. This, of course, is the underlying reason for the validity of adiabatic theory as a kinematical foundation for the study of radiation-belt dynamics.

I.3 Liouville's Theorem

The kinematical state of a particle in three-dimensional motion can, in general, be defined instantaneously by specifying its three coordinates of position and three components of canonical momentum. These six quantitites locate a point in the six-dimensional continuum known as *phase space*. As time evolves, the motion of the particle generates a trajectory in phase space.

A system consisting of N distinct particles of a given species (*e. g.*, protons) is described by a set of N distinct points in phase space. When N is very large, it proves convenient to describe the distribution of these points in phase space by means of a six-dimensional density function $f(\pi_i, q_i; t)$ where the π_i $(i=1,2,3)$ are components of canonical momentum, the q_i are coordinates of position, and t is the time. This distribution function has the usual significance that $f d^3\pi \, d^3\mathbf{q}$ is the number of particles instantaneously occupying the six-dimensional volume $d^3\pi \, d^3\mathbf{q}$ in canonical phase space.

In Hamiltonian mechanics the temporal evolution of $f(\pi_i, q_i; t)$ is specified by Liouville's theorem [7], which asserts that

$$(df/dt) \equiv (\partial f/\partial t) + \sum_{i=1}^{3} [\dot{\pi}_i(\partial f/\partial \pi_i) + \dot{q}_i(\partial f/\partial q_i)] = 0 \qquad (1.08)$$

along any dynamical trajectory in phase space. In more picturesque terms, the phase-space volume containing the system of N distinct representative points moves incompressibly through phase space. The dynamical trajectory of an individual particle is governed by the equations

$$\dot{\pi}_i = -\partial H/\partial q_i \qquad (1.09\,a)$$

$$\dot{q}_i = \partial H/\partial \pi_i, \qquad (1.09\,b)$$

where $H(\pi_i, q_i; t)$ is the Hamiltonian. Liouville's theorem implies that the content of an infinitesimal six-dimensional volume $d^3\pi\, d^3q$ surrounding a particle's location in phase space remains invariant as the particle executes its dynamical trajectory. This means that a unit Jacobian characterizes the transformation of such an infinitesimal volume moving in accordance with the laws of classical mechanics. Indeed, the execution of a dynamical trajectory is describable by a sequence of infinitesimal contact transformations [7]. Each of these infinitesimal transformations has the properties that

$$\pi_i \rightarrow \pi_i + d\pi_i = \pi_i - (\partial H/\partial q_i)dt \qquad (1.10\,a)$$

$$q_i \rightarrow q_i + dq_i = q_i + (\partial H/\partial \pi_i)dt \qquad (1.10\,b)$$

$$H \rightarrow H + dH = H + (\partial H/\partial t)dt \qquad (1.10\,c)$$

in accordance with (1.09).

Apart from its utility in specifying the adiabatically invariant action integrals J_i, the canonical momentum π is not a convenient physical quantity in the study of radiation-belt dynamics. It is far more convenient to deal with the locally observable particle momentum $\mathbf{p}$ alone than in combination with $(q/c)\mathbf{A}$, since the electromagnetic vector potential $\mathbf{A}$ is neither locally observable nor uniquely defined. Accordingly, it becomes advantageous at this point to introduce the distribution function $f(\mathbf{p}, \mathbf{r}; t)$, which represents the density of particles in a six-dimensional (but non-canonical) position-momentum space. The relation between $f(\mathbf{p}, \mathbf{r}; t)$ and $f(\boldsymbol{\pi}, \mathbf{q}; t)$ is readily obtained via the algebraic transformation

$$\boldsymbol{\pi} = \mathbf{p} + (q/c)\mathbf{A} \qquad (1.11\,a)$$

$$\mathbf{q} = \mathbf{r}. \qquad (1.11\,b)$$

No loss of generality is suffered by supposing that the coordinates are Cartesian. In this case it is easy to verify that the transformation defined by (1.11) has a unit Jacobian, from which it follows that $f(\mathbf{p}, \mathbf{r}; t) = f(\boldsymbol{\pi}, \mathbf{q}; t)$. The distribution function $f(\mathbf{p}, \mathbf{r}; t)$, of course, has the significance that $f d^3\mathbf{p}\, d^3\mathbf{r}$ is the number of particles instantaneously occupying the infinitesimal six-dimensional volume $d^3\mathbf{p}\, d^3\mathbf{r}$ in position-momentum space. Since $f(\mathbf{p}, \mathbf{r}; t)$ is numerically equal to the phase-space density $f(\boldsymbol{\pi}, \mathbf{q}; t)$, it follows that f remains constant along a dynamical trajectory in position-momentum space. This property is summarized by the equation (known as the Vlasov equation)

$$(\partial f/\partial t)+(\mathbf{p}/m)\cdot(\partial f/\partial \mathbf{r})+\mathbf{F}\cdot(\partial f/\partial \mathbf{p})=0, \tag{1.12}$$

where $\mathbf{F}=\dot{\mathbf{p}}$ is the force applied to a particle of momentum $\mathbf{p}$ located at position $\mathbf{r}$. The particle velocity $\mathbf{v}$ is equal to $\mathbf{p}/m$, and the relativistic mass m exceeds the rest mass m_0 by the factor

$$\gamma=[1+(p/m_0 c)^2]^{1/2}, \tag{1.13}$$

where c is the speed of light.

The Vlasov equation is sometimes cast in the alternative form [18]

$$(\partial f/\partial t)+\sum_i v_i(\partial f/\partial r_i)+\sum_{ij}(F_i/m)[\delta_{ij}-(v_i v_j/c^2)](\partial f/\partial v_j)=0, \tag{1.14}$$

where δ_{ij} is the Kronecker symbol ($=1$ for $i=j$ and $=0$ for $i \neq j$). This form can be derived from (1.12) by noting that

$$m(\partial v_j/\partial p_i)=\delta_{ij}-(v_i v_j/c^2). \tag{1.15}$$

As formulated here, the Vlasov equation takes account of the relativistic kinematics of charged particles but does not include certain processes (such as collisions) that are not easily described by a Hamiltonian. Such processes are best added phenomenologically to the Vlasov description.

I.4 The Dipole Field

For purposes of analytical calculation it is often convenient to represent the geomagnetic field as the field of a magnetic dipole centered within a perfectly spherical earth. The dipole axis is assumed to be coincident with the axis of rotation, and the spherical polar coordinates r, θ, and φ are measured from the center of the earth, the north pole, and the midnight meridian, respectively. The field intensity is given by

$$\mathbf{B}=-B_0(2\hat{\mathbf{r}}\cos\theta+\hat{\boldsymbol{\theta}}\sin\theta)(a/r)^3, \tag{1.16}$$

where a is the radius of the earth and $B_0(\approx 0.31$ gauss$)$ is the equatorial $(\theta=\pi/2)$ magnitude of $\mathbf{B}$ at $r=a$. A field line that intersects the equatorial plane at a distance $r=La$ from the origin generates a drift shell to which is assigned the dimensionless parameter L. The differential equation of this field line is

$$dr/d\theta=r B_r/B_\theta=2r\operatorname{ctn}\theta, \tag{1.17a}$$

from which it is deduced that

$$r=La\sin^2\theta. \tag{1.17b}$$

The element of arc length along the field line is therefore

$$ds = La(1 + 3\cos^2\theta)^{1/2}\sin\theta\,d\theta, \tag{1.18}$$

and from this expression follows the value given by (1.01) for the total length S of the field line.

In the formal theory of adiabatic motion it is customary to introduce the Euler potentials α and β such that $\mathbf{A} = \alpha\nabla\beta$ and (therefore) $\mathbf{B} = \nabla\alpha \times \nabla\beta$. The dipole field can be generated by assigning the Euler potentials $\alpha = -B_0 a^2/L$ and $\beta = \varphi$, so that

$$\mathbf{A} = -B_0\,\hat{\boldsymbol{\varphi}}\,(a^3/r^2)\sin\theta. \tag{1.19}$$

These assignments are not unique, since they can be modified by a gauge transformation without altering any physical consequences. The chosen representations, however, have the special significance that $|\Phi| = -2\pi\alpha$ and $\varphi_3 = \beta$, i.e., the Euler potentials are immediately related to the third invariant and drift phase, respectively. Euler potentials are used elsewhere in describing the geomagnetic field [19], but are omitted from further discussion in this volume.

The kinematical description of the radiation belts is simplified greatly by the customary assumption that field lines are equipotential. Of course, this assumption is not rigorously justified, but for particles of sufficiently large energy the variations of time-independent electrostatic and gravitational potentials along a field line are unimportant[4].

The adiabatic motion of a charged particle influenced only by a magnetostatic field of mirror geometry conserves both M, as given by (1.05), and the kinetic energy $E = m_0 c^2(\gamma - 1)$. It follows from (1.13), then, that $p^2(\equiv p_\perp^2 + p_\parallel^2)$ remains constant, where $p_\parallel$ is the component of momentum parallel to $\mathbf{B}$. This component of $\mathbf{p}$ vanishes at each mirror point, where the guiding-center magnetic-field intensity is denoted B_m. Here the angle between $\mathbf{p}$ and $\mathbf{B}$, known as the *pitch angle*, is $90°$. The minimum angle between $\mathbf{p}$ and $\mathbf{B}$ attained during the bounce period is known as the *equatorial pitch angle*, because this minimum occurs at that point along the guiding field line at which B is a minimum; in the dipole field this point lies on the equatorial plane.

The bounce motion of a particle's guiding center along an equipotential field line has a period

$$2\pi/\Omega_2 = (m/p)\oint[1 - (B/B_m)]^{-1/2}\,ds, \tag{1.20}$$

---

[4]Gravity is totally negligible for particle energies $\gtrsim 1\,\mathrm{keV}$. Electrostatic-potential variations along a field line in the radiation zone may amount to 10—100 volts, and so are similarly negligible for particle energies $\gtrsim 1\,\mathrm{keV}$ [11].

where the integral is evaluated along the guiding field line. Both p and M are constants of the bounce motion, and the integral above can be interpreted either as twice the *spiral* path length between mirror points along the actual trajectory, or as the integral of the pitch-angle secant along the guiding-center trajectory. The instantaneous value of $\dot{s}$, in other words, is given by

$$p_{\parallel}/m=[(p/m)^2-(2m_0MB/m^2)]^{1/2}=(p/m)[1-(B/B_m)]^{1/2}. \quad (1.21)$$

The maximum value of $p_{\parallel}$ along the bounce path corresponds to the minimum of B. Thus, if x is the cosine of the equatorial pitch angle and B_e is the equatorial guiding-center field magnitude, it follows that

$$x^2=1-(B_e/B_m)=1-y^2, \quad (1.22)$$

where y is the sine of the equatorial pitch angle.

For the dipole field it follows from (1.16) and (1.17 b) that $B_e=B_0L^{-3}$ and that

$$B=(B_0/L^3)(1+3\cos^2\theta)^{1/2}\csc^6\theta. \quad (1.23)$$

The bounce period $2\pi/\Omega_2$ is therefore representable as

$$2\pi/\Omega_2=(4mLa/p)T(y), \quad \tfrac{p}{m}=(x^2-1)^{1/2}c \quad (1.24a)$$

where

$$T(y)=\int_{\theta_m}^{\pi/2}\frac{\sin\theta(1+3\cos^2\theta)^{1/2}d\theta}{[1-y^2\csc^6\theta(1+3\cos^2\theta)^{1/2}]^{1/2}}. \quad (1.24b)$$

The colatitude θ_m of the northern mirror point is given by the relation

$$y=(1+3\cos^2\theta_m)^{-1/4}\sin^3\theta_m. \quad (1.25)$$

At $y=0$ the integral for $T(y)$ is easily evaluated. The result is $T(0)=S/2La$, where S (the length of the field line) is given by (1.01). At $y=1$ $(\theta_m=\pi/2)$ the integral can be evaluated by an appeal to the theory of small-amplitude oscillations [7] about the equator. The equation of motion for such a particle subjected to magnetostatic forces is

$$m\ddot{s}=-(M/\gamma)(\partial B/\partial s)=-(p^2/2mB_e)(\partial^2B/\partial s^2)_es, \quad (1.26)$$

where the subscript e denotes the equator $(s=0)$. The magnetic moment in general is equal to M/γ, and this amounts to $p^2/2mB_e$ for an equatorially mirroring particle. The equatorial value of $\partial^2B/\partial s^2$ is given by

$$(\partial^2B/\partial s^2)_e=\hat{\mathbf{B}}_e\cdot\nabla(\hat{\mathbf{B}}\cdot\nabla B)_e=(3/La)^2B_e, \quad (1.27)$$

so that $\Omega_2=(p/m)(3/\sqrt{2}La)$. It follows that $T(1)=(\pi/6)\sqrt{2}$.

For $0 < y < 1$, exact evaluation of (1.24b) in terms of elementary functions of y is impossible. A very good estimate, however, is provided by the formula [20]

$$T(y) \approx T(0) - \tfrac{1}{2}[T(0) - T(1)](y + y^{1/2}),\qquad(1.28\,\text{a})$$

where

$$T(0) = 1 + (1/2\sqrt{3})\ln(2 + \sqrt{3}) \approx 1.3802\qquad(1.28\,\text{b})$$

$$T(1) = (\pi/6)\sqrt{2} \approx 0.7405\qquad(1.28\,\text{c})$$

$$\tfrac{1}{2}[T(0) - T(1)] \approx 0.3198 ;\qquad(1.28\,\text{d})$$

at worst, this estimate deviates from the numerically computed function $T(y)$ by less than 1 % (see Table 1).

Table 1. Functions of Bounce Motion in Dipole Field

θ_m	$y^{1/2}$	$\sin^{-1} y$	Exact T	Approx T	Exact Y	Approx Y
0°	0.00000	0.00°	1.380	1.380	2.760	2.760
1°	0.00194	0.00°	1.380	1.380	2.760	2.758
5°	0.02165	0.03°	1.376	1.373	2.741	2.730
10°	0.06102	0.21°	1.366	1.359	2.682	2.6633
15°	0.1114	0.71°	1.350	1.341	2.587	2.565
20°	0.1701	1.66°	1.327	1.316	2.457	2.434
25°	0.2352	3.17°	1.298	1.287	2.296	2.275
30°	0.3051	5.34°	1.264	1.253	2.109	2.091
35°	0.3785	8.23°	1.224	1.213	1.901	1.886
40°	0.4539	11.89°	1.179	1.169	1.678	1.666
45°	0.5303	16.33°	1.129	1.121	1.446	1.437
50°	0.6062	21.56°	1.076	1.069	1.211	1.205
55°	0.6804	27.58°	1.020	1.014	0.9793	0.9761
60°	0.7515	34.38°	0.963	0.959	0.7577	0.7562
65°	0.8178	41.97°	0.906	0.905	0.5521	0.5517
70°	0.8773	50.32°	0.854	0.853	0.3693	0.3692
74°	0.9186	57.54°	0.816	0.816	0.2438	0.2438
78°	0.9528	65.20°	0.784	0.785	0.1408	0.1408
82°	0.9785	73.23°	0.760	0.761	0.06386	0.06387
86°	0.9945	81.54°	0.745	0.746	0.01617	0.01617
90°	1.0000	90.00°	0.740	0.740	0.00000	0.00000

The second invariant J, as given by (1.06), can be approximated by means of a formula derivable from (1.28). The invariant is given by

$$J = 2pLaY(y),\qquad(1.29\,\text{a})$$

where

$$Y(y)=2\int_{\theta_m}^{\pi/2}\frac{\sin\theta(1+3\cos^2\theta)^{1/2}d\theta}{[1-y^2\csc^6\theta(1+3\cos^2\theta)^{1/2}]^{-1/2}}.\qquad(1.29\,\mathrm{b})$$

The observation that

$$\frac{d}{dy}\left(\frac{Y}{y}\right)=-\frac{2\,T}{y^2}\qquad(1.30)$$

enables $Y(y)$ to be estimated from (2.28). Since $Y(1)=0$, it follows that

$$Y(y)=2y\int_y^1 u^{-2}\,T(u)\,du\approx2(1-y)\,T(0)+[T(0)-T(1)](y\ln y+2\,y-2\,y^{1/2}).\qquad(1.31)$$

This estimate remains within 1% of the numerically computed $Y(y)$ for all values of y between 0 and 1 (see Table 1). Moreover, the exact analytical result that $Y(0)=2\,T(0)$ is reproduced by (1.31). An expansion for $x^2\equiv1-y^2\ll1$ reproduces the harmonic-oscillator approximation, which implies that

$$J=\oint p_{||}ds=\oint(p_{||}^2/m)dt\approx(p^2x^2/2m)(2\pi/\Omega_2)\qquad(1.32\mathrm{a})$$

or

$$Y(y)\approx T(1)x^2=(\pi/6)\sqrt{2}x^2\approx0.7405(1-y^2).\qquad(1.32\mathrm{b})$$

In (1.32a) the time-averaged value of $p_{||}^2$ is equal to half the maximum value, since $p_{||}$ is a harmonically varying quantity in the limit of vanishing bounce amplitude. This maximum value, attained at the equator, is p^2x^2.

As a further application of (1.28) it is possible to estimate the pitch-angle dependence of the azimuthal drift frequency $\Omega_3/2\pi$. According to the sign convention introduced above, the drift phase φ_3 is a temporally increasing quantity, so that $\dot\varphi_3=-(q/|q|)\Omega_3$, where $\Omega_3=\dot\varphi$ is the time derivative of the particle's azimuthal coordinate (see Sections I.1 and I.2). It follows from Hamilton-Jacobi theory [7] that

$$\Omega_3=-(2\pi q/|q|)(\partial H/\partial J_3)_{M,J}\qquad(1.33)$$

where H $(=\gamma m_0 c^2)$ is the Hamiltonian. Evaluation of this expression is facilitated by noting that (1.13) implies $dH/dp=p/m$, while (1.07) implies $J_3=|q|(2\pi B_0 a^2/cL)$. It follows from (1.05), (1.29), and (1.30) that

$$(\partial\ln y/\partial\ln L)_{M,J}=-Y/4\,T\qquad(1.34\mathrm{a})$$

and

$$(\partial p/\partial L)_{M,J}=(p/4LT)(Y-6T)\equiv-[3p/LT(y)]D(y), \quad (1.34\,\mathrm{b})$$

where

$$12D(y)=6T(y)-Y(y). \quad (1.34\,\mathrm{c})$$

Simple algebraic manipulations thus lead to the formula

$$\Omega_3/2\pi = -(3\gamma L/2\pi)(p/ma)^2(m_0c/qB_0)[D(y)/T(y)]$$
$$= -(3L/2\pi\gamma)(\gamma^2-1)(c/a)^2(m_0c/qB_0)[D(y)/T(y)] \quad (1.35)$$

for the azimuthal drift frequency.

Since $Y(1)=0$ and $Y(0)=2T(0)$, it follows at once that $T(1)=2D(1)$ and $T(0)=3D(0)$. For intermediate values of y, an accurate analytical approximation to $D(y)$ is provided by (1.28) and (1.31). Explicitly stated, the result is

$$12D(y)\approx4T(0)-[3T(0)-5T(1)]y-[T(0)-T(1)](y\ln y+y^{1/2}). \quad (1.36)$$

The estimate for $D(y)/T(y)$ provided by (1.28) and (1.36) deviates at

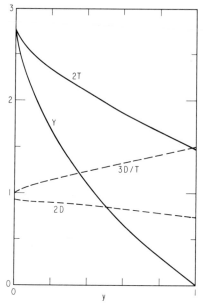

Fig. 7. Functions governing the pitch-angle dependence of bounce and drift frequencies in a dipole field.

worst by less than 0.2% from the numerically computed ratio $D(y)/T(y)$. Figure 7 indicates graphically the dependence of $T(y)$, $Y(y)$, $D(y)$, and $D(y)/T(y)$ upon y. The functions $D(y)$ and $T(y)$ and their ratio all vary monotonically with y in such a manner that, for a given particle species and energy, the bounce frequency and drift frequency are maximal for particles that mirror at the equator and progressively smaller for particles having mirror points at progressively higher (poleward) latitudes. This variation of bounce and drift frequency with equatorial pitch angle ($\sin^{-1} y$), however, is remarkably weak, as it amounts to less than a factor of two in each case.

The equations of this section summarize the adiabatic motion of a particle trapped in the field of a magnetic dipole centered within the earth. The three invariants M, J, and Φ are conserved, and (since the field is symmetric in azimuth) the drift shell is generated by rotating the guiding field line about the dipole axis. The gyrofrequency $\Omega_1/2\pi$ varies with the instantaneous position of the guiding center along the field line. The minimum magnitude of Ω_1 is given by $|q|B_0/mcL^3$ and is attained at the equator. The maximum value is $|q|B_0/mcL^3y^2$ and is attained at the mirror latitude. Since the drift shell is symmetric about the dipole axis, the bounce frequency $\Omega_2/2\pi$ is independent of azimuth and given by $\Omega_2/2\pi = [p/4mLaT(y)]$. A further consequence of azimuthal symmetry is that the bounce-averaged azimuthal coordinate φ advances eastward (in the case of a negatively charged particle) or retreats westward (for a positive ion) at a constant rate equal to the value of Ω_3 given by (1.35). The functional forms of $T(y)$, $Y(y)$, and $D(y)$ noted above are, of course, valid only for the dipole field.

I.5 The Distorted Field

The centered-dipole field is only a gross idealization of the true geomagnetic field. The idealized field has value as a standard of reference for the analysis of radiation-belt dynamics, however, and this value is enhanced by an appreciation of the extent to which the true field deviates from the ideal.

The dipolar component of the earth's field originates in the molten core. Higher multipoles of the core field diminish in intensity by comparison with the dipolar component and are relatively unimportant at geocentric distances of order one earth radius and beyond. Measurements made at the earth's surface, however, suggest that the dipole axis is tilted 11.4° relative to the rotation axis and displaced ~400 km from the center of the earth[5]. True magnetic anomalies (deviations from a dipolar field) can originate either from higher multipoles of the core field or from concentrations of ferromagnetic material in the earth's crust. In addition, currents can be induced in the earth and in the ionosphere by virtue of the earth's rotation and by externally produced disturbances (see below) of the geomagnetic field. These induced currents can be very complicated in structure; fortunately, they are not known to have a dominant influence on the radiation belts.

The principal distortions of the earth's *outer* magnetosphere are caused by currents on the magnetopause, on the neutral sheet, and within the magnetosphere itself. The current layer that constitutes the magnetopause serves to confine the geomagnetic field within. Thus, the magnetopause is a boundary beyond which the earth's field does not extend. The neutral sheet, which separates the oppositely oriented flux tubes that constitute the geomagnetic tail, carries currents that tend generally to weaken the nightside field intensity. Together, the magnetopause and neutral sheet form the magnetospheric surface.

The final source of field distortion important for the radiation belts is the *ring current* carried by the hot component of the magnetospheric plasma. The direction of gradient-curvature drift in the earth's field is westward for protons and eastward for electrons, and indeed the *net* ring current flows *westward*. The result is a generally outward displacement of field lines, *i. e.*, a decreased magnetic-field intensity at the earth's surface and elsewhere interior to the ring-current zone, but an enhanced field strength at exterior points. With particle gyration taken into account, the spatial distribution of electric-current *density* is found to have a more subtle structure than consideration of gradient-curvature drift alone would suggest. For reasons discussed below, the local current density actually is directed eastward at the inner edge of the ring-current belt, but the net current carried by a spatially bounded hot plasma does flow westward, in accordance with the unsophisticated expectation.

It is generally considered impractical to model all the aforementioned current systems simultaneously and self-consistently. In studying radiation-belt dynamics by theoretical means, however, it is usually sufficient to recognize that self-consistent magnetic-field models exist in principle. Thus, it is possible to *imagine* the computation of a particle's three

---

[5]The 400-km displacement causes the apogee of an inner-zone particle drift shell to be located over the western Pacific Ocean. Conversely, perigee is attained over the south Atlantic. Since the field intensity at a given geocentric *altitude* over the south Atlantic is substantially smaller than at other geographic locations, this region where drift shells attain perigee is often called the *South Atlantic* "*anomaly*".

adiabatic invariants and three phases with the understanding that these invariants and phases identify an equivalent "particle" trapped in the centered-dipole reference field. In particular, this mental exercise assigns to the particle a unique, significant, and adiabatically invariant shell parameter [5]

$$L = 2\pi a^2 B_0 |\Phi|^{-1}, \tag{1.37}$$

i.e., the shell parameter of the adiabatically equivalent "particle" in the dipole field.

The McIlwain Parameter. In most cases it is, in fact, necessary to carry out some form of adiabatic transformation of *observational* data so as to establish a requisite degree of order. In practice, quiet-time observations of the inner radiation zone ($L \lesssim 3$) need be corrected only for anomalies of the permanent geomagnetic field (including displacement of the point dipole). Since this field is constant over the lifetime of a typical satellite experiment, it is customary to circumvent computation of the invariant shell parameter given by (1.37). It is found that observational data from the inner zone can be ordered adequately by specifying the mirror field B_m and the second invariant J for particles of known energy. It is customary to derive from these quantities a non-invariant shell parameter L_m, defined as the dipole shell parameter of a "particle" having the same B_m, J, and energy. To facilitate the calculation of L_m, it is usual to introduce the quantity $I \equiv J/2p$. In a dipole field it is found that

$$y^2 (I^3 B_m/a^3 B_0) \equiv y^2 R = [Y(y)]^3 \tag{1.38}$$

where $y^2 = B_0/L^3 B_m$. The shell parameter L_m is thus defined by the relation $L_m^3 = B_0/y^2 B_m$, where y is the solution of (1.38). The value of I is computed within the framework of an empirical model of the permanent geomagnetic field [6], and the value of $R (\equiv I^3 B_m/a^3 B_0)$ is thereby determined [21].

Since the function $Y(y)$ given by (1.29) cannot be expressed in closed form by elementary functions, an exact algebraic solution of (1.38) for y is impossible to obtain. Moreover, the analytical approximation to $Y(y)$ given by (1.31) does not render (1.38) algebraically tractable as an equation to be solved for y. A numerical solution is possible, of

---

[6]The parameter L_m originally defined by McIlwain [21] is computed by assigning $B_0 = 0.311653$ gauss. However, it would be more reasonable to compute L_m using the best available field model and the corresponding value of B_0 for the epoch in question [22].

course, but for most purposes the empirically deduced relationship [22]

$$B_m L_m^3 / B_0 \approx 1 + (18/\pi^2)^{1/2} R^{1/3} + 0.465380 R^{2/3} + [Y(0)]^{-3} R$$
$$\approx 1 + 1.350474 R^{1/3} + 0.465380 R^{2/3} + 0.047546 R \qquad (1.39)$$

is entirely adequate for defining L_m in terms of I and B_m. Indeed, the error in using (1.39) to specify L_m amounts to less than 0.01%, as compared with the value of L_m obtained via the exact dipole function given by (1.29 b). The coordinates B_m and L_m, generally called (B, L) coordinates, are known to order inner-zone particle data satisfactorily during magnetically quiet periods, in spite of certain conceptual difficulties; *e.g.*, the fact that (even in an azimuthally symmetric field) particles having the same Φ can be assigned different values of L_m. The utility of (B, L) space for describing the inner zone during quiet periods resides in the fact that such conceptual discrepancies are of insignificant magnitude there. For example, the variation of L_m (as computed from a standard 512-term multipole expansion of the permanent geomagnetic field) among particles whose mirror points lie along a given field line amounts consistently to less than 1% [21].

The Ring Current. During magnetically disturbed periods it is necessary to take adiabatic account of ring-current effects in both the inner ($L \lesssim 3$) and outer ($L \gtrsim 3$) radiation zones. As a very crude approximation, the ring current may be compared to a solenoid located beyond $L \approx 3$. This approximation suggests a roughly uniform field perturbation oriented parallel to the dipole axis and extending throughout the inner zone[7]. This perturbing field often attains a magnitude $\gtrsim 200\gamma$ during magnetic storms and is closely associated with the equatorial geomagnetic index D_{st}, which is supposed to measure the azimuthally symmetric component of the axial field perturbation induced by the storm [23]. The signature of D_{st} (as obtained from low-latitude magnetograms) is usually negative because the perturbing field points southward, thereby diminishing the field intensity at the earth's surface, where the equatorial dipolar component points generally northward. (On exceptional occasions, when the ring current is weak, a positive D_{st} can result from compression of the magnetosphere by the solar wind.)

During a magnetic storm the ring current tends to coexist spatially with a portion of the outer zone, and so it would be a poor approximation to extend the uniform perturbing field beyond $L \approx 3$. Indeed, within

---

[7]Temporal changes in the uniform axial field induce (via surface currents) an effective magnetic dipole in the earth, which is essentially a perfect conductor on the time scale of a geomagnetic storm.

the belt of ring-current protons and electrons, diamagnetic effects can accentuate the field depression beyond that seen at the earth's surface ($r = a$). On the outer fringes of the ring-current belt, the field depression is greatly reduced. At sufficiently large distances the ring current would resemble a magnetic dipole (of finite extent) aligned with the earth's dipole. The result is therefore an augmentation of the earth's field at such distances. Since self-constistent models of the ring current and its magnetic field [24] require considerable computation, it is customary to employ semi-empirical models to account for the associated adiabatic effects upon radiation-belt particles.

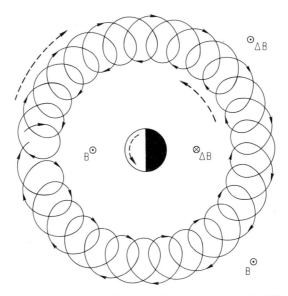

Fig. 8. Schematic representation of the gyration and azimuthal drift (solid curve) of an equatorially mirroring proton, with associated current patterns (dashed curves).

Figure 8 illustrates the drift-phase averaged current pattern associated with the gyration and gradient drift of an equatorially mirroring proton. The current pattern in this case has a width of two gyroradii. The inner portion of this pattern carries an eastward current, while the outer portion carries a somewhat larger westward current. The net flux of electrical current across each meridional half-plane is westward, as provided by the gradient drift. Formulation of a ring-current model consists of superimposing the contributions of all protons and electrons in the hot plasma, whose behavior is governed by the self-consistent field.

In a plasma having pressure $P_\perp$ in the direction normal to **B**, the *magnetization current* (caused by particle gyration) has a density $\mathbf{J}_m = -c\nabla \times (P_\perp \mathbf{B}/B^2)$. Under nonrelativistic conditions, gradient drift produces a current density $\mathbf{J}_g = c(P_\perp/B^2)\hat{\mathbf{B}} \times \nabla B$, and curvature drift yields a current density $\mathbf{J}_c = c(P_\parallel/B^2)\mathbf{B} \times (\partial\hat{\mathbf{B}}/\partial s)$, where $P_\parallel$ is the component of *sign ?* plasma pressure parallel to **B**. The magnetization current can be written in the expanded form

$$\mathbf{J}_m = -c(P_\perp/B^2)\nabla \times \mathbf{B} + (c/B^2)\mathbf{B} \times \nabla P_\perp - 2\mathbf{J}_g \qquad (1.40a)$$

and the gradient-drift current can be expressed as

$$\mathbf{J}_g = c(P_\perp/B^2)[\mathbf{B} \times (\partial\hat{\mathbf{B}}/\partial s) - (\nabla \times \mathbf{B})_\perp]. \qquad (1.40b)$$

In differential geometry the normal vector $\partial\hat{\mathbf{B}}/\partial s$ has a magnitude equal to the local curvature of the field line and points toward the center of curvature. Under the static conditions considered here, the total current density **J** satisfies the relation $c\nabla \times \mathbf{B} = 4\pi\mathbf{J} = 4\pi(\mathbf{J}_m + \mathbf{J}_g + \mathbf{J}_c)$ and is given [25] by

$$\mathbf{J} = (c/B)\hat{\mathbf{B}} \times \nabla P_\perp + (c/8\pi)(\beta_\parallel - \beta_\perp)\mathbf{B} \times (\partial\hat{\mathbf{B}}/\partial s), \qquad (1.41)$$

where $\beta_\parallel = 8\pi P_\parallel/B^2$ and $\beta_\perp = 8\pi P_\perp/B^2$. The beta parameters relate the pressures exerted by the hot plasma to that exerted by the magnetic field. Observations of the earth's ring current indicate that $\beta_\parallel$ and $\beta_\perp$ both attain magnitudes of order unity in the region of space most densely populated by protons in the energy range 10—50 keV [15]. This region lies in the vicinity of $L=3$ during large magnetic storms and near $L=7$ during geomagnetically quiet times[8].

The initial term of (1.41) points in the eastward $(+\hat{\boldsymbol{\varphi}})$ direction in the inner portion of the ring-current zone, but in the westward $(-\hat{\boldsymbol{\varphi}})$ direction in the outer portion. Since $\hat{\mathbf{B}} \times \nabla P_\perp$ is weighted by $1/B$ in (1.41), the westward contribution predominates if the hot plasma is spatially bounded[9].

Simplified models of the ring-current field [26] can be constructed empirically, by allowing the field perturbation to have a fixed spatial profile whose amplitude is directly proportional to D_{st}. Such a model is illustrated in Fig. 9, where $\Delta\mathbf{B}$ is the *equatorial* **B**-field perturbation caused by the ring current. As noted above, this total field perturbation

---

[8]High-beta conditions also characterize the vicinity of the dayside neutral points and the nightside neutral sheet. Elsewhere in the magnetosphere it is found that both $\beta_\parallel$ and $\beta_\perp$ are rather small in comparison with unity.

[9]Similarly, the inner edge of the plasma sheet can carry an eastward current, even though the predominant flow of current on the neutral sheet is westward, in accordance with the expectation based on gradient drift.

includes the earth-induction field $\mathbf{B}_i$ which can be simulated by placing a point dipole at the geocenter. The ratio of B_i to the field of the earth's permanent dipole is very small, *i. e.*, approximately the ratio of D_{st} to one gauss. The form of $\Delta \mathbf{B}$ beyond $L \sim 3$ is not really independent of D_{st}, as the model implies. In fact, the diamagnetic field depression resides at $L \lesssim 4$ only when $|D_{st}| \gtrsim 100 \gamma$. The region of maximum hot-plasma energy density is observed to be correlated with D_{st} in such a manner that beta attains a value of order unity there [15]. Thus, the diamagnetic depression moves outward in L with decreasing $|D_{st}|$. However, the ring current exerts a negligible influence on the radiation belts when D_{st} is smaller than $\sim 30 \gamma$ in absolute value. The model summarized by Fig. 9 is therefore adequate in the sense that an accurate profile of $\Delta B/D_{st}$ is needed only for the rather large values of $|D_{st}|$ to which Fig. 9 applies.

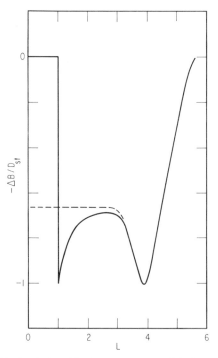

Fig. 9. Semi-empirical relationship between equatorial ring-current field ΔB and magnetic index D_{st} [26], including effects of currents induced on the surface of a perfectly conducting earth (solid curve); with such earth-induction field sub-tracted out (dashed curve).

The Mead Field. In addition to ring-current effects, the earth's magnetic field is permanently distorted by virtue of currents that flow on the magnetopause and neutral sheet. Various models are available for describing the effects of these currents in a quantitative manner [27, 28]. The permanent compression of the magnetosphere caused directly by the solar wind can best be evaluated by expanding a scalar potential function $V_m(r, \theta, \varphi)$ in spherical harmonics and deriving from this potential the magnetic field $\mathbf{B} = -\nabla V_m$. The coefficients introduced in the spherical-harmonic expansion are then evaluated by requiring pressure balance at the resulting magnetopause. This procedure is greatly simplified by supposing that the earth's dipole is normal to the direction of the undeflected solar wind. The use of a scalar potential $V_m(r, \theta, \varphi)$ implies the neglect of plasma-pressure effects (e. g., currents) within the magnetosphere. Pressure balance at the magnetopause therefore requires that $B^2 = 8\pi\rho_s u^2(1 - \cos\psi_s)$ at each point on this boundary, where ρ_s is the mass density of solar-wind material, $\mathbf{u}$ is the velocity of the undeflected solar wind, and ψ_s is the angle of deflection caused by encounter with the magnetopause. This formulation ignores the interplanetary magnetic field, whose energy density is smaller than that of the flowing plasma by a factor ~ 100.

A simplified picture of solar-wind deflection by the magnetopause postulates specular reflection of the plasma. In this case the angle ψ_s is twice the local angle of attack of the incident solar wind. The resulting coefficients $\bar{g}_l^m$ in the expansion

$$V_m(r, \theta, \varphi) = -B_0(a^3/r^2)\cos\theta + (a^3/b^2)\sum_{l=1}^{\infty}(r/b)^l\,\bar{g}_l^m P_l^m(\cos\theta)\cos m\varphi\,, \quad (1.42)$$

which exhibits north-south and dawn-dusk symmetry by virtue of the assumed orthogonality of $\mathbf{u}$ to the dipole axis, define the Mead field [28]. The symbol $P_l^m(\cos\theta)$ denotes an associated Legendre polynomial with Schmidt normalization [10], as is conventional for geomagnetic applications. The computed values of $\bar{g}_l^m/B_0$ are given in Table 2, and b is the equatorial "stand-off" distance from the point dipole to the magnetopause in the noon meridian. From the indicated coefficients it follows that

$$b = 1.068 \; B_0^2/4\pi\rho_s u^2)^{1/6}a, \quad (1.43)$$

so that $b \approx 10a$ under typical solar-wind conditions.

---

[10]The functions $P_l^m(x)$ are defined by the equations

$$P_l^m(x) = \left[\frac{2(l-m)!}{(l+m)!}\right]^{1/2}\frac{(1-x^2)^{m/2}}{2^l\,l!}\frac{d^{l+m}}{dx^{l+m}}[(x^2-1)^l], \quad m>0$$

$$P_l^m(x) = \frac{1}{2^l\,l!}\frac{d^l}{dx^l}[(x^2-1)^l], \quad m=0.$$

Table 2. Expansion Coefficients for Mead Field

l,m	$\bar{g}_l^m/B_0$	l,m	$\bar{g}_l^m/B_0$
1,0	0.8100	5,0	0.0184
2,1	0.4065	5,2	−0.0348
3,0	−0.0233	5,4	−0.0032
3,2	−0.0752	6,1	−0.0042
4,1	0.0775	6,3	0.0061
4,3	0.0052	6,5	0.0013

For illustrative purposes the potential given by (1.42) can be simplified further by neglecting those coefficients $\bar{g}_l^m$ that have $l > 2$. The simplified potential can then be written [28] in the form

$$V_m(r, \theta, \varphi) = -B_0(a^3/r^2)\cos\theta - [B_1 z - B_2 z(x/b)](a/b)^3, \quad (1.44)$$

where $x = r\sin\theta\cos\varphi$, $z = r\cos\theta$, $B_0 \approx 0.31$ gauss, $B_1 = -\bar{g}_1^1 \approx 0.25$ gauss, and $B_2 = \sqrt{3}\bar{g}_2^1 \approx 0.21$ gauss. Very often the simplified field derived from (1.44) can be utilized fruitfully in analytical calculations related to adiabatic motion and particle diffusion. In polar coordinates this simplified field has the form

$$B_r = -2B_0(a/r)^3\cos\theta + B_1(a/b)^3\cos\theta$$
$$\quad - 2B_2(a/b)^4(r/a)\cos\theta\sin\theta\cos\varphi \quad (1.45\,a)$$

$$B_\theta = -B_0(a/r)^3\sin\theta - B_1(a/b)^3\sin\theta$$
$$\quad + B_2(a/b)^4(r/a)(2\sin^2\theta - 1)\cos\varphi \quad (1.45\,b)$$

$$B_\varphi = B_2(a/b)^4(r/a)\cos\theta\sin\varphi, \quad (1.45\,c)$$

where r is the geocentric distance, θ is the colatitude measured from the northern pole, and φ is the east longitude measured from the midnight meridian. Figure 10 illustrates field-line traces for this model and the dipole field in the plane for which $\sin\varphi = 0$. For this purpose field lines are identified by the label L_d, defined as the limit of $(r/a\sin^2\theta)$ as θ approaches zero. This definition is motivated by (1.17 b).

The simplified Mead field given by (1.45) can be considered a special case of the general analytic representation [29]

$$B_r(r, \theta, \varphi; t) = \sum_{lmn} B_r(l, m, n; t)(r/a)^n \cos\theta\sin^l\theta\cos m\varphi \quad (1.46\,a)$$

$$B_\theta(r, \theta, \varphi; t) = \sum_{lmn} B_\theta(l, m, n; t)(r/a)^n \sin^l\theta\cos m\varphi \quad (1.46\,b)$$

$$B_\varphi(r, \theta, \varphi; t) = \sum_{lmn} B_\varphi(l, m, n; t)(r/a)^n \cos\theta\sin^l\theta\sin m\varphi \quad (1.46\,c)$$

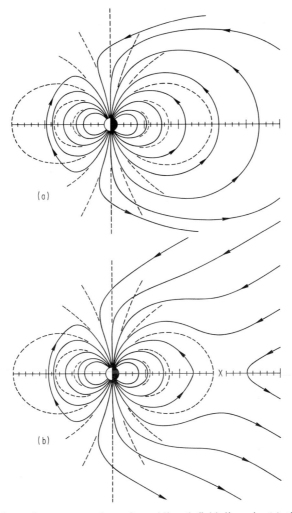

Fig. 10. Schematic representation of meridional field lines in (a) the 13-term and (b) the 3-term magnetospheres (solid curves). Corresponding dipole field lines (dashed curves) are shown for $\Lambda = 65°$, $70°$, $75°$, $80°$, $85°$, and $90°$, but omitted for $\Lambda = 60°$, where $\Lambda \equiv \sec^{-1}(L_d^{1/2})$. The symbol X marks the location of the nightside neutral line that automatically appears in the 3-term model.

of a magnetic field having symmetry with respect to the equatorial plane and the noon-midnight meridional plane. The completeness of (1.46) as the expansion of an analytic function having these symmetries is quite evident. The radial variable r enters in the form of a Taylor-Laurent series. Functions of the colatitude θ that are even with respect to $\theta = \pi/2$ can surely be written as a power series in $\sin\theta$. As for functions that are odd with respect to $\theta = \pi/2$, the factor $\cos\theta$ need enter only to the first power, since even powers of $\cos\theta$ can be written as polynomials in $\sin\theta$. Finally, a Fourier series in $\sin m\varphi$ or $\cos m\varphi$ will suffice to express an analytic function that is odd or even with respect to the midnight meridian ($\varphi = 0$).

Since (1.46) is considerably more general than a spherical-harmonic expansion of $V_m(r, \theta, \varphi; t)$, it can be used even in the presence of ring currents and other distributed sources. All that is required for this extension is that care be taken to satisfy the relation $c\nabla \times \mathbf{B} = 4\pi\mathbf{J} + (\partial\mathbf{E}/\partial t)$, where $\mathbf{E}$ is the electric field and $\mathbf{J}$ is the current density. In addition to this requirement, of course, the magnetic field must be made to satisfy $\nabla \cdot \mathbf{B} = 0$ under all conditions. This general requirement leads to the constraining equation

$$(n+2)B_r(l-1, m, n; t) + (l+1)B_\theta(l, m, n; t) + m B_\varphi(l, m, n; t) = 0, \quad (1.47)$$

which indeed is satisfied by the coefficients

$$B_r(0, 0, -3) = 2B_\theta(1, 0, -3) = -2B_0 \qquad (1.48\,\text{a})$$

$$B_r(0, 0, 0) = -B_\theta(1, 0, 0) = B_1(a/b)^3 \qquad (1.48\,\text{b})$$

$$B_r(1, 1, 1) = 2B_\theta(0, 1, 1) = -B_\theta(2, 1, 1)$$
$$= -2B_\varphi(0, 1, 1) = -2B_2(a/b)^4 \qquad (1.48\,\text{c})$$

$$B_0 = 1.24\,B_1 = 1.48\,B_2 = 0.31 \text{ gauss} \qquad (1.48\,\text{d})$$

appropriate to the simplified Mead model given by (1.45). As noted above, the simplified field is especially useful for carrying out analytical calculations appropriate to a model magnetosphere. In some cases however, the simplified model is not accurate enough to organize observational data obtained beyond $L \approx 5$, *i.e.*, to recast such data in terms of a standard magnetosphere by means of the required adiabatic transformations.

The Mead-Williams Field. The usual shortcoming of (1.45) in the description of observational data is the neglect of the neutral-sheet currents associated with the geomagnetic tail. In principle, these currents fit into the framework of (1.46) so long as they are distributed in space rather than confined to an idealized sheet of vanishing thickness. In other words, so long as $\mathbf{J}$ is everywhere finite in magnitude, there are no singularities in $\mathbf{B}$ that (1.46) fails to handle. In practice, however, it is customary to represent the neutral sheet in the idealized manner. This means that the Mead field (typically as derived from Table 2) is augmented by the field of a current-carrying sheet located on the nightside equator. Various representations are possible. The popular Mead-Williams field [30] represents the current sheet as a strip of finite width ($x=x_n$ to $x=x_f$, subscripts denoting the near and far boundaries) extending from $y=-\infty$ to $y=+\infty$ (see Fig. 3). The current is assumed to be distributed uniformly between $x=x_n$ and $x=x_f$ and flows from $y=+\infty$ (east) to $y=-\infty$ (west). Near the current sheet itself, the resulting tail field $\mathbf{B}_t$ has a magnitude

$$B_t=(2\pi/c)(x_f-x_n)^{-1}I,\qquad (1.49)$$

where I is the total current carried. This field points sunward $(-\hat{\mathbf{x}})$ in the northern hemisphere and antisunward $(+\hat{\mathbf{x}})$ in the southern hemisphere. As a result, nightside polar field lines (i.e., those emanating from the earth at polar latitudes) are greatly extended in the equatorial region.

   With the aid of a system of synchronous satellites ($r=6.6a$, $\dot{\varphi}=2\pi/$day) it becomes possible to compile a magnetospheric "weather report" providing both b (the stand-off distance) and B_t (the tail field) as functions of time. The method is to compare the observed magnetometer readings at various longitudes with those predicted by assuming various combinations of the two model parameters. The determination of b and B_t is made most confidently by comparing equatorial values of B at $\varphi=0$ (midnight) and $\varphi=\pi$ (noon). The model parameters are then defined by locating B_e ($\varphi=0$) and B_e ($\varphi=\pi$) in Fig. 11, which is a contour plot of b and B_t [31]. In case multiple-satellite coverage is not available at the synchronous orbit, it may be necessary to utilize

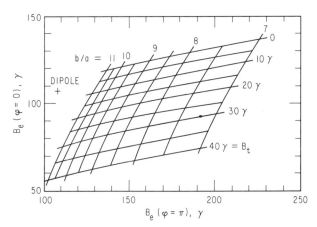

Fig. 11. Equatorial magnitudes of $\mathbf{B}$ at noon $(\varphi=\pi)$ and midnight $(\varphi=0)$ in synchronous orbit ($r=6.6a$) computed [31] for selected values of the Mead-Williams parameters b/a and B_t [30], with $x_n/a=10.6-0.06(B_t/1\gamma)$ and $x_f/a=200$.

readings of a single magnetometer taken at twelve-hour intervals for the purpose outlined above. This procedure is acceptable as long as magnetospheric conditions do not change significantly during the twelve-hour interval between midnight and noon [31]. Using the report of the day-to-day variation of b and B_t obtained by this method, it is possible to recast particle data obtained beyond $L \approx 3$ in terms of a standard magnetosphere, e.g., the dipole field. In this case each particle in the distorted field is to be identified with an equivalent "particle" having the same three adiabatic invariants and phases (but probably a different pitch angle and energy) in the dipole field. For $L \lesssim 5$ it is usually unnecessary to make such a transformation, as the effects of currents on the magnetopause and neutral sheet are small in magnitude there, when compared with the magnitude of the dipole field (see Section 1.4).

There exists some doubt that the subsolar point $(r, \theta, \varphi) = (b, \pi/2, \pi)$ on the magnetopause should be treated as a point of specular reflection (as in the Mead model) rather than a point of hydrodynamic stagnation [10]. The truth presumably lies somewhere between these two limits. If ρ_s is the density of mass flowing at the solar-wind velocity $\mathbf{u}$, pressure balance at the subsolar point is expressed by the relation $\rho_s u^2 = B^2/4\pi$ for specular reflection and by $\rho_s u^2 = B^2/8\pi$ for hydrodynamic stagnation. The superficial consequence of the uncertainty involved here is a possible error $\sim 12\%$ in specifying b by means of (1.43); a deeper consideration of the hydrodynamic model would require a difficult recomputation of the coefficients $\bar{g}_l^m$ that appear in Table 2. Existing problems in the field of radiation-belt dynamics, however, appear to transcend such subtleties in modeling the magnetosphere.

Similarly, a more realistic model of the tail field might take into account the fact that the relevant current loops close over the cylindrical surface of the magnetosphere rather than at infinity. By thus restricting the current sheet to a lateral dimension $\lesssim 4b$, it is possible to extend x_f to infinity, or at least to realistically large distances ($\gtrsim 50b$) without catastrophe to the dayside magnetosphere. An additional element of reality would be introduced by taking account of the tilt that exists between the earth's dipole and the solar wind. Since the dipole axis is inclined $11.4°$ to the rotation axis, which in turn is inclined $22.5°$ to the ecliptic plane, the tilt of the dipole away from normal solar-wind incidence can amount to as much as $34°$, depending on time of day and season of year. In recent years considerable progress has been made toward constructing models that account for the influence of tilt on the shape of the magnetopause, the character of the distorted field, and the position of the neutral sheet. As with the questions raised in the paragraph above, it appears that these considerations are quite important in defining the overall structure of the magnetosphere (including, for example, the description of diurnal and seasonal variations characteristic of ground-based magnetometer readings), but that these complicating effects are not of crucial importance to the radiation belts *per se*. In other words, the existing state of knowledge concerning the earth's trapped-radiation environment does not justify the additional labor inherent in more realistically describing the containing magnetic field for radiation-belt studies. Accordingly, the field models employed in subsequent analyses will be kept as simple as possible.

I.6 Magnetospheric Electric Fields

Large-scale electric fields in the magnetosphere originate primarily from temporal variations of the magnetic field, from the rotation of the earth, and from plasma instabilities of the neutral sheet. Electric fields induced by temporal variations of $\mathbf{B}$ are not derivable from an electrostatic (scalar) potential V_e, because $c\nabla \times \mathbf{E} = -\partial \mathbf{B}/\partial t$. Those resulting from the earth's rotation and from neutral-sheet instabilities can be derived from scalar potentials. The electrostatic field caused by the rotation of a magnetic dipole about its axis, taken as an idealized geophysical situation, is given by

$$\mathbf{E} = -(1/c)(\mathbf{\Omega}_0 \times \mathbf{r}) \times \mathbf{B}$$
$$= B_0(\Omega_0 a/c)(2\,\hat{\boldsymbol{\theta}}\cos\theta - \hat{\mathbf{r}}\sin\theta)(a/r)^2 \sin\theta \qquad (1.50)$$

where $\mathbf{\Omega}_0$ is the angular velocity of the earth. This field can be derived from the potential

$$V_e(r,\theta,\varphi) = -B_0\Omega_0 a^2/c\,L_d, \qquad (1.51\,\text{a})$$

where

$$L_d \equiv \lim_{\theta\to 0}(r/a\sin^2\theta). \qquad (1.51\,\text{b})$$

The limit indicated in (1.51 b) must be evaluated along the magnetic field line. The field-line label L_d proves to be useful in other applications in which internal geomagnetic multipoles are neglected, *e.g.*, those involving the Mead field, hence the need for a precise definition. A dipole field line, of course, identically satisfies the relation $r = L_d a \sin^2\theta$.

The so-called convection electric field $\mathbf{E}_c$ required to maintain the tail (neutral-sheet) current in the presence of intrinsic plasma turbulence is customarily represented via the potential

$$V_e(r,\theta,\varphi) = E_c a L_d \sin\varphi \qquad (1.52)$$

in the region where $\mathbf{B}$ is dipolar. This expression reduces to $V_e(r,\pi/2,\varphi)$ $=E_c y$ in the equatorial plane, where the idealized field attains the spatially uniform (in two dimensions) magnitude E_c, directed from dawn to dusk $(-\hat{\mathbf{y}})$. This field drives a sunward $(-\hat{\mathbf{x}})$ convection of plasma in the forward portion of the magnetosphere, where $\hat{\mathbf{B}}=\hat{\mathbf{z}}$ in the equatorial plane. The plasma flow follows from the standard relation

$$\mathbf{v}_d=(c/B^2)\mathbf{E}\times\mathbf{B}. \tag{1.53}$$

In the magnetotail, however, the general direction of $\mathbf{B}$ is either $-\hat{\mathbf{x}}$ (northern hemisphere; $z>0$) or $+\hat{\mathbf{x}}$ (southern hemisphere; $z<0$). The result is a plasma flow velocity

$$\mathbf{v}_d=-\hat{\mathbf{z}}(c\,E_c/B_t)(z/|z|) \tag{1.54}$$

directed into the neutral sheet.

A picturesque interpretation of (1.54) is that the field lines themselves flow into the neutral sheet at a speed $c\,E_c/B_t$, there experiencing a mutual annihilation that liberates energy at a rate of $2(B_t^2/8\pi)(c\,E_c/B_t)$ per unit area since $B_t^2/8\pi$ is the density of field energy. Similarly, it is possible to view sunward convection of plasma as a "snapping back" of field lines that have been dragged downstream by a viscous interaction with the solar wind. Indeed, there exists such a viscous interaction at the magnetopause, but it acts fundamentally upon the *plasma* rather than upon the *field* [12]. Plasma and field-line motion can be identified in terms of (1.53) by requiring also that $\mathbf{E}=-(1/c)\mathbf{v}_d\times\mathbf{B}$ [see (1.50)]. There exists, then, a choice between postulating field-line motion at velocity $\mathbf{v}_d$ accompanied by plasma motion at the same velocity (line tying) on the one hand, and the convection of plasma at velocity $\mathbf{v}_d$ across a stationary $\mathbf{B}$ field on the other.

Particularly in the steady state, for which $\partial\mathbf{B}/\partial t=0$, the dual concepts of field-line motion and line tying can be very confusing when taken as a foundation for quantitative analysis. The description based on physically measurable quantities such as $\mathbf{E}$ and $\mathbf{B}$ is never less adequate than the more colorful description, and usually yields more readily to quantification.

The electrostatic fields derived from (1.51) and (1.52) have the property that $\mathbf{E}\cdot\mathbf{B}=0$, where $\mathbf{B}$ is given by (1.16). The property $\mathbf{E}\cdot\mathbf{B}=0$ seems to be essential for the identification of field-line motion with cold-plasma motion [32]. Accordingly, it has become conventional to postulate the condition $\mathbf{E}\cdot\mathbf{B}=0$ as a means of mapping magnetospheric electric fields. The usual rationale for this postulate is that the magnetosphere contains cold plasma of sufficient density to short out any appreciable field-aligned component of $\mathbf{E}$ [11][11]. Applicability of the condition $\mathbf{E} \cdot \mathbf{B} = 0$ anywhere beyond the plasmasphere is a matter of some controversy, although (as noted) this procedure is the conventional one for mapping magnetospheric electric fields [33].

It is inappropriate to employ (1.51) and (1.52) where the magnetic field differs significantly from that of a dipole. One prescription for obtaining $\mathbf{E}$ in the distorted field involves an expansion analogous to (1.46). This prescription defines an iterative procedure [34] whereby the condition $\mathbf{E} \cdot \mathbf{B} = 0$ is imposed order by order in r/b, beginning with (1.50).

Operating within the framework of the dipole field, it is not difficult to establish the existence of both closed and open equipotential surfaces of the superimposed convection and corotation electric fields. The total electrostatic potential of this idealized steady-state magnetosphere has the form

$$V_e(r, \theta, \varphi) = E_c L_d a \sin \varphi - B_0 (\Omega_0 a^2 / c L_d), \tag{1.55}$$

and so equipotential (constant-V_e) surfaces are specified by

$$L_d = (2 E_c \, a \sin \varphi)^{-1} \{ V_e \pm [V_e^2 + (4 E_c B_0 \Omega_0 a^3 / c) \sin \varphi]^{1/2} \}. \tag{1.56}$$

Examples are illustrated in Fig. 12; the singular equipotential surface that separates the closed and open cold-plasma drift shells is that for which $V_e = -2(E_c B_0 \Omega_0 a^3 / c)^{1/2}$. This shell, which satisfies the equation

$$L_d = (B_0 \Omega_0 a / c E_c)^{1/2} [(1 + \sin \varphi)^{1/2} - 1] \csc \varphi, \tag{1.57}$$

is closely associated with a virtual discontinuity in the magnetospheric cold-plasma density. The underlying reason for this *plasmapause* is that ionospheric plasma originating at low and middle latitudes remains trapped within closed equipotential surfaces, while that originating at sufficiently high latitudes proceeds to escape from the magnetosphere.

The dimensionless parameter $(B_0 \Omega_0 a / c E_c)^{1/2}$ appearing in (1.57) measures (in earth radii) the nominal radius of the plasmasphere. More precisely, this parameter identifies the equatorial geocentric distance to the plasmapause at the dusk meridian ($\varphi = -\pi/2$), which in this idealized model corresponds to the "bulge" region, *i.e.*, the region of maximum geocentric radius. A plasmasphere *diameter* of six earth radii in the noon-midnight meridian corresponds to a *radius* of six earth radii at dusk in this model, and leads to the estimate that $E_c \sim 4 \, \mu\text{V/cm}$

---

[11]Beyond the radiation belts, *e.g.*, in the auroral zone, violations of this rule are quite common. The auroral zone appears to be associated with the earthward portion of the plasma sheet.

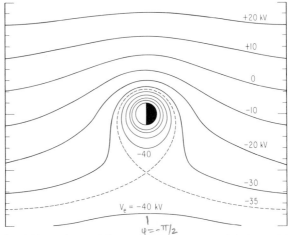

Fig. 12. Electrostatic equipotential contours in the equatorial plane of the idealized geomagnetic dipole, computed for $E_c = 0.526\,\text{V/km}$ (contours are unlabeled for $V_e < -40\,\text{kV}$ and omitted for $V_e < -70\,\text{kV}$).

under typical conditions. In reality, the plasmasphere exhibits a somewhat less pronounced azimuthal asymmetry, and the "bulge" appears roughly midway through the evening quadrant ($\varphi \sim -\pi/4$). This means that (1.52) somewhat oversimplifies the actual convection electric field. Moreover, the size of the plasmasphere [see (1.57)] is found empirically to vary with magnetic activity (e.g., with the geomagnetic index K_p) in a manner compatible with the statistical relationship [35]

$$E_c \approx 5.65\,(m_e c^2/q_p a)(u/c), \tag{1.58}$$

where m_e is the mass of an electron, q_p is the charge of a proton, and u is the solar-wind speed. This formula yields $E_c \approx 6\,\mu\text{V/cm}$ for $u = 400\,\text{km/sec}$. A statistical correlation between E_c and u is intuitively appealing in that sunward convection of plasma is supposed to balance (on average) the outward flow characteristic of a viscous boundary layer at the magnetopause [12].

The steady component of the magnetospheric electric field imposes a final restriction on radiation-belt particle energies. Convention requires that gradient-curvature drifts dominate adiabatic $\mathbf{E} \times \mathbf{B}$ drifts, at least to the point of guaranteeing existence of the third invariant; i.e., a closed drift shell (see Introduction). Preferably, the particle energy should be large enough that the drift shell deviates insignificantly from that calculated in the absence of magnetospheric electric fields. This condition imposes the requirement that $E/L \gg |q|\,E_c a \sim 4\,\text{keV}$ (see above) and demarcates the outer radiation belt from the ring-current belt, with which it is spatially coincident. Within the plasmasphere, however, it is necessary to operate in a frame rotating with the earth unless $EL \gg |q| B_0 \Omega_0 a^2/c$ $\sim 100\,\text{keV}$ [see (1.51)]. Particles not satisfying this criterion should probably be excluded from consideration, since the usual radiation-belt methods and scaling laws (e. g., Fig. 6) do not apply without this modification. Figure 13 summarizes the parametric demarcations that distinguish radiation-belt particles from the other inhabitants of the earth's magnetosphere, based on the various considerations outlined in the present chapter.

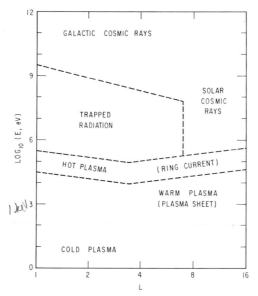

Fig. 13. Spatial and spectral classification of charged particles in the magnetosphere.

I.7 Flux Mapping and Shell Tracing

As a general rule, the particle flux $J_\alpha(E;\mathbf{r})$ per unit energy E per unit solid angle Ω at local pitch angle α is related to the distribution function $f(p_{\parallel}, p_{\perp}; \mathbf{r})$ of (1.12) by the formula

$$J_\alpha(E;\mathbf{r})\, dE\, d\Omega = f(p_{\parallel}, p_{\perp}; \mathbf{r})(p/m) p^2 \, dp\, d\Omega, \qquad (1.59)$$

where $m = \gamma m_0$ is the relativistic mass and m_0 is the rest mass. The total energy $mc^2 = E + m_0 c^2$ is related to the scalar momentum p by the equation

$$(E + m_0 c^2)^2 = p^2 c^2 + m_0^2 c^4 = p_{\parallel}^2 c^2 + p_{\perp}^2 c^2 + m_0^2 c^4, \qquad (1.60)$$

where $p_{||} = p \cos \alpha$ and $p_{\perp} = p \sin \alpha$. From (1.60) it follows that $m\, dE = p\, dp$, and so (1.59) becomes

$$J_{\alpha}(E;\mathbf{r}) = p^2 f(p_{||}, p_{\perp}; \mathbf{r}). \tag{1.61}$$

This equation is specialized to the case of locally mirroring particles by requiring $p_{||} = 0$ ($\alpha = \pi/2$), so that

$$J_{\perp}(E;\mathbf{r}) = 2 m_0 M B f(0, p_{\perp}; \mathbf{r}). \tag{1.62}$$

Since Liouville's theorem (Section I.3) assures that $f(p_{||}, p_{\perp}; \mathbf{r})$ remains constant along a dynamical trajectory in phase space, conservation of the first adiabatic invariant means that $J_{\perp}/B$ at constant M remains fixed along the trajectory of a representative particle's mirror point in either hemisphere.

In an azimuthally symmetric magnetosphere, the tracing of drift shells would be very simple. Each shell could be generated by rotating a field line about the axis of symmetry. Particles mirroring at different latitudes along a given field line would proceed to generate coincident drift shells in the course of adiabatic motion, and the equatorial pitch angle of each particle would remain unchanged with azimuthal drift.

In the earth's magnetosphere, this azimuthal degeneracy is broken not only by the day-night asymmetry of the **B** field, as represented by (1.45), but also (to a lesser extent for radiation-belt particles) by the dawn-dusk asymmetry of the convection electric field, as represented by (1.52). As a consequence, the drift shells generated by the adiabatic motion of particles identical in species and energy, sharing a common field line at some longitude, generally do not coincide at other longitudes if the particles have different pitch angles at the equator of the common field line. This adiabatic phenomenon is known as *shell splitting* [5]. The extent to which drift shells are split by the azimuthal asymmetries can be judged by independently tracing the shells that correspond to distinct equatorial pitch angles $\sin^{-1} y$ on the common field line. It is instructive to consider the two dominant asymmetries separately.

Electric Shell Splitting. When the electrostatic potential given above by (1.55) is superimposed on the magnetic-dipole field (1.16), the tracing of drift shells is accomplished by employing conservation of $J = 2 L_d a \cdot p\, Y(y)$, $M = p^2/2 m_0 B_m$, and $J^2/8 m_0 a^2 M = B_m L_d^2 Y^2$, where $B_m = B_0/L_d^3 y^2$. The conserved energy $W \equiv E + q V_e(\mathbf{r})$ is given by

$$2 m_0 c^2 M B_m + m_0^2 c^4 = (m_0 c^2 + W - q V_e)^2. \tag{1.63}$$

From these identities follow the differential relationships

$$(1/B_m)(d B_m/d\varphi) = -(2/L_d)(d L_d/d\varphi) - (2/Y)(d y/d\varphi) Y'(y), \tag{1.64a}$$

$$(2/y)(d y/d\varphi) = -(3/L_d)(d L_d/d\varphi) - (1/B_m)(d B_m/d\varphi), \tag{1.64b}$$

and

$$m_0 c^2 M(d B_m/d \varphi) = \tag{1.64 c}$$
$$- q(m_0 c^2 + W - q V_e)[(\partial V_e/\partial L_d)_\varphi (d L_d/d \varphi) + (\partial V_e/\partial \varphi)_{L_d}].$$

Taken together, equations (1.64) and (1.30) yield the drift-shell equations

$$d L_d/d \varphi = L_d T(y)(m_0 c^2 + W - q V_e)[3(W - q V_e)(2 m_0 c^2 + W - q V_e) D(y)$$
$$- T(y)(m_0 c^2 + W - q V_e) q(\partial V_e/\partial L_d)_\varphi]^{-1} q(\partial V_e/\partial \varphi)_{L_d} \tag{1.65 a}$$

$$d y/d \varphi = -(y/4 L_d)[Y(y)/T(y)](d L_d/d \varphi). \tag{1.65 b}$$

Thus, evolution of the drift trajectory $L_d(\varphi)$ clearly depends upon y, and so the nonvanishing of $(\partial V_e/\partial \varphi)_{L_d} = E_c a L_d \cos \varphi$ leads to the splitting of drift shells. In the limit of very weak shell splitting ($|q V_e| \ll W$), the lowest-order approximation

$$L_d(\varphi) \approx L_d(0)\{1 + [(m_0 c^2 + W)/(2 m_0 c^2 + W)]$$
$$\times (q E_c a L_d/3 W)[T(y)/D(y)] \sin \varphi\} \tag{1.66}$$

follows from (1.65). The ratio T/D is a monotonically decreasing function of y.

Magnetic Shell Splitting. For sufficiently small absolute values of the expansion parameter $q E_c a L_d/W$, the shell splitting predicted by (1.66) is negligible compared with that caused by azimuthal asymmetry of the *magnetic* field. To evaluate this latter effect, it is proper to neglect $V_e(\mathbf{r})$ and calculate the energy-independent drift shells imposed by (1.45). In this case the expansion parameter $\varepsilon_2 \equiv (B_2/B_0)(L_d a/b)^4 \ll 1$ characterizes the azimuthal asymmetry. With the neglect of $V_e(\mathbf{r})$, the variables p and B_m become constants of the motion in a static $\mathbf{B}$ field. It is necessary, however, to generalize from the dipole functions $T(y)$ and $Y(y)$ so that $2\pi/\Omega_2$ and J can be written

$$2\pi/\Omega_2 = 4 L_d a(m/p) \tilde{T}(y; L_d, \varphi) \tag{1.67 a}$$

$$J = 2 L_d a p \tilde{Y}(y; L_d, \varphi). \tag{1.67 b}$$

The derivation leading to (1.30) equally well relates $\tilde{Y}(y; L_d, \varphi)$ to $\tilde{T}(y; L_d, \varphi)$. Generalization of (1.28) to the non-dipolar $\mathbf{B}$ field (1.45) requires at least that $\tilde{T}(0; L_d, \varphi)$ and $\tilde{T}(1; L_d, \varphi)$ be calculated to lowest order in $\varepsilon_1 \equiv (B_1/B_0)(L_d a/b)^3$ and $\varepsilon_2 \equiv (B_2/B_0)(L_d a/b)^4$. The results are given by

$$\tilde{T}(0; L_d, \varphi) = 1 + (I_1/2) + \varepsilon_1(15 I_9 - 16 I_7)$$
$$- \varepsilon_2(147 I_{12} - 205 I_{10} + 54 I_8)\cos\varphi \qquad (1.68\,\text{a})$$

$$\tilde{T}(1; L_d, \varphi) = (\pi/6)\sqrt{2}[1 + \varepsilon_1 - (25\,\varepsilon_2/14)\cos\varphi], \qquad (1.68\,\text{b})$$

where

$$I_n \equiv \int\limits_0^{\pi/2} \frac{\sin^n\theta\,d\theta}{(1 + 3\cos^2\theta)^{1/2}}, \qquad (1.68\,\text{c})$$

and follow from a tracing of field lines that deviate from the dipole solution $r = L_d a \sin^2\theta$. To lowest order in ε_1 and ε_2, the distorted field lines satisfy the equations

$$d\ln r/d\theta = B_r/B_\theta = 2\,\text{ctn}\,\theta - 3\varepsilon_1\sin^5\theta\cos\theta$$
$$+ 2\,\varepsilon_2(3\sin^2\theta - 1)\sin^6\theta\cos\theta\cos\varphi \qquad (1.69\,\text{a})$$

$$r = L_d a \sin^2\theta\big[1 - (\varepsilon_1/2)\sin^6\theta$$
$$+ (2\varepsilon_2/21)(7\sin^2\theta - 3)\sin^7\theta\cos\varphi\big]. \qquad (1.69\,\text{b})$$

The functional value of $\tilde{T}(0; L_d, \varphi)$ is defined by the requirement that $L_d a \tilde{T}(0; L_d, \varphi)$ be equal to half the arc length of the entire field line, i.e.,

$$L_d a \tilde{T}(0; L_d, \varphi) = \int\limits_0^{\pi/2} [r^2 + (dr/d\theta)^2 + r^2\sin^2\theta(d\varphi/d\theta)^2]^{1/2}d\theta. \quad (1.70)$$

Since $(d\varphi/d\theta)^2$ is of higher order than first in ε_2, this contribution to the arc length is neglected in obtaining (1.68 a) directly from (1.69) and (1.70). Numerical values of I_n are listed in Table 3, together with the specific combinations needed in (1.68 a).

The derivation of (1.68 b) follows that of (1.28 c). The harmonic bounce approximation requires that $(\partial^2 B/\partial s^2)$ defined as $\hat{\mathbf{B}} \cdot \nabla(\hat{\mathbf{B}} \cdot \nabla B)$, be eval-

Table 3. Selected Integrals I_n

$I_1 = 0.760346$	$I_7 = 0.406500$	$I_{13} = 0.315466$
$I_2 = 0.630306$	$I_8 = 0.385465$	$I_{14} = 0.305646$
$I_3 = 0.553737$	$I_9 = 0.367590$	$I_{15} = 0.296713$
$I_4 = 0.501251$	$I_{10} = 0.352080$	$I_{16} = 0.288537$
$I_5 = 0.462142$	$I_{11} = 0.338446$	$I_{17} = 0.281016$
$I_6 = 0.431423$	$I_{12} = 0.326330$	$I_{18} = 0.274066$

$$1 + (1/2)I_1 = 1.380173$$
$$15 I_9 - 16 I_7 = -0.988542$$
$$147 I_{12} - 205 I_{10} + 54 I_8 = -3.390834$$

uated on the magnetic-equatorial surface, *i. e.*, on the surface for which $\partial B/\partial s=0$ and $\partial^2 B/\partial s^2>0$. For $\mathbf{B}$ given by (1.45) and $r \lesssim 0.8b$ this surface coincides with the plane $\theta=\pi/2$. A general expression for $\partial^2 B/\partial s^2$ at $\theta=\pi/2$, applicable whenever $\mathbf{B}$ is derivable from a scalar potential and has north-south symmetry about this plane, is

$$Br^2(\partial^2 B/\partial s^2)=2r^2(\partial B/\partial r)^2+3B(\partial B/\partial r)+B^2$$
$$+2(\partial B/\partial \varphi)^2+B_\theta(\partial^2 B_\theta/\partial \theta^2). \tag{1.71}$$

In particular, if $\mathbf{B}$ is given by (1.45), the result is

$$Br^2(\partial^2 B/\partial s^2)=9B^2-9B_1(a/b)^3[3B-2B_1(a/b)^3]$$
$$+2B_2^2(a/b)^6(r/b)^2(1+15\cos^2\varphi) \tag{1.72}$$
$$+B_2(a/b)^3(r/b)[39B-48B_1(a/b)^3]\cos\varphi$$

at $\theta=\pi/2$. Taken to lowest order ε_1 and ε_2, this result combines with (1.69b) to yield

$$B^{-1/2}(\partial^2 B/\partial s^2)^{1/2}=(3/L_d a)[1-\varepsilon_1+(25\,\varepsilon_2/14)\cos\varphi]. \tag{1.73}$$

Since $\Omega_2^2=(p^2/2m^2 B)(\partial^2 B/\partial s^2)$ according to (1.26), it follows from (1.67a) and (1.73) that $\tilde{T}(1;L_d,\varphi)$ is correctly specified by (1.68b).

It is consistent with (1.30) to express the functions $\tilde{T}(y;L_d,\varphi)$ and $\tilde{Y}(y;L_d,\varphi)$ in the form

$$\tilde{T}(y;L_d,\varphi)=\tilde{T}(0;L_d,\varphi)+\varepsilon_1 y^2 G_1'(y)+\varepsilon_2 y^2 G_2'(y)\cos\varphi$$
$$-(1/2)[\tilde{T}(0;L_d,\varphi)-\tilde{T}(1;L_d,\varphi)](y+y^{1/2}) \tag{1.74a}$$

$$\tilde{Y}(y;L_d,\varphi)=(2+y\ln y-2y^{1/2})\tilde{T}(0;L_d,\varphi)$$
$$-(2y+y\ln y-2y^{1/2})\tilde{T}(1;L_d,\varphi) \tag{1.74b}$$
$$-2\varepsilon_1 yG_1(y)-2\varepsilon_2 yG_2(y)\cos\varphi,$$

where $G_1(1)=G_1'(1)=G_2(1)=G_2'(1)=0$. Except for these four end-point constraints dictated by (1.68), the functions $G_1(y)$ and $G_2(y)$ remain to be specified (see below).

The drift-shell equation for constant p and B_m $(=B_e/y^2)$ is obtained by invoking the constancy of J [given by (1.67b)]; it follows that

$$dL_d/d\varphi=-[(\partial B_e/\partial \varphi)_{L_d}(\tilde{Y}-2\tilde{T})+2B_e(\partial \tilde{Y}/\partial \varphi)_{L_d}] \tag{1.75}$$
$$\div[2(B_e/L_d)\tilde{Y}+2B_e(\partial \tilde{Y}/\partial L_d)_\varphi+(\partial B_e/\partial L_d)_\varphi(\tilde{Y}-2\tilde{T})],$$

where B_e is the field intensity at $\theta=\pi/2$. From (1.45b) and (1.69b) it follows that

$$B_e=(B_0/L_d^3)[1+(5\varepsilon_1/2)-(15\varepsilon_2/7)\cos\varphi]. \tag{1.76}$$

If (1.75) is then evaluated to lowest order in the field asymmetry, the result is

$$d L_d/d \varphi = - [L_d^4/12 B_0 D(y)][2(B_0/L_d^3)(\partial \tilde{Y}/\partial \varphi)_{L_d} \\ + (15 B_2/7)(a/b)^4 (Y - 2 T) L_d \sin \varphi] \qquad (1.77\,a)$$

$$d L_d/d \varphi \approx - (B_2/21 B_0) L_d^5 (a/b)^4 (1/12 D)[45(Y - 2 T) \\ + 4(147 I_{12} - 205 I_{10} + 54 I_8)(2 + y \ln y - 2 y^{1/2}) \qquad (1.77\,b) \\ + 42 y G_2(y) - 75 T(1)(2 y + y \ln y - 2 y^{1/2})] \sin \varphi ,$$

where (1.34 c) defines $12 D(y) \equiv 6 T(y) - Y(y)$. The use of (1.77 b) is made convenient by the development of an analytical approximation for

$$Q(y) \equiv 45 Y(y) - 90 T(y) + 42 y G_2(y) - 75 T(1)(2 y + y \ln y - 2 y^{1/2}) \\ + 4(147 I_{12} - 205 I_{10} + 54 I_8)(2 + y \ln y - 2 y^{1/2}), \qquad (1.78\,a)$$

which must satisfy the conditions

$$Q(1) = - 90 T(1) \approx - 66.6432441 \qquad (1.78\,b)$$

$$Q(0) = 8(147 I_{12} - 205 I_{10} + 54 I_8) \approx - 27.1266694 \qquad (1.78\,c)$$

$$Q'(1) = (15/2)[9 T(0) - 41 T(1)] \approx - 134.5360732 . \qquad (1.78\,d)$$

Exact numerical evaluations [19, 36] of $T(y)$, $Y(y)$, and the shell-splitting function $Q/12 D$ yield the functional values $Q(y)$ given in Table 4. Plotted on a graph (not shown here), these functional values indicate that $Q(y)$ varies only weakly with y for $y \lesssim 0.4$, but quite strongly for $y \gtrsim 0.7$, and that $Q(y)$ is *almost* a monotonic function. The empirical representation [66]

$$Q(y) \approx Q(0) + [2Q(1) - 2Q(0) - (1/4)Q'(1)] y^4 \\ + [Q(0) - Q(1) + (1/4)Q'(1)] y^8 \qquad (1.79) \\ \approx - 27.12667 - 45.39913 y^4 + 5.88256 y^8$$

provides numerical accuracy well within 1% over the entire range of y (see Table 4) in addition to satisfying the end-point ($y = 0$ and $y = 1$) requirements exactly. It is therefore proper to use (1.79) in combination with (1.28), (1.31), and (1.36) in the tracing of magnetospheric drift shells. No approximation for $G_1(y)$ is needed for this purpose, and so none has been developed.

Integration of (1.77 b) with respect to φ leads directly to a lowest-order expression for tracing drift shells whose pitch-angle degeneracy is broken

Table 4. Exact and Approximate Values of $Q(y)$

θ_m	y	Exact Q	Approx. Q	$\Delta Q/Q$
$0°$	0.000000	-27.127	-27.127	0.0000
$20°$	0.028947	-27.085	-27.127	$+0.0015$
$30°$	0.093098	-27.090	-27.130	$+0.0015$
$40°$	0.206042	-27.118	-27.208	$+0.0033$
$50°$	0.367471	-27.777	-27.953	$+0.0063$
$55°$	0.462962	-29.051	-29.200	$+0.0051$
$60°$	0.564719	-31.645	-31.683	$+0.0012$
$65°$	0.668717	-36.026	-35.970	-0.0016
$70°$	0.769660	-42.509	-42.333	-0.0041
$75°$	0.860893	-50.378	-50.289	-0.0018
$80°$	0.934656	-58.313	-58.347	$+0.0006$
$85°$	0.983074	-64.323	-64.398	$+0.0012$
$90°$	1.000000	-66.643	-66.643	0.0000

by the day-night asymmetry of **B**. Two limiting cases of notable simplicity are recovered from (1.77b). For $y=1$ the drift trajectory is a path of constant B on the equatorial surface. For $y=0$ the drift shell follows field lines of equal arc length.

Numerical evaluation of (1.72) for the reasonable values $b=10a$ and $B_0=1.24B_1=1.48B_2=0.31$ gauss reveals that the right-hand side becomes negative on the day side ($\cos\varphi<0$) for $r\gtrsim 8a$. This behavior signals a bifurcation of the equatorial (minimum-B) surface as one approaches the magnetopause. In other words, dayside field lines for which L_d is sufficiently large ($\gtrsim 10$) satisfy $\partial B/\partial s=0$ and $\partial^2 B/\partial s^2>0$ at points symmetrically displaced in magnetic latitude from the equatorial plane of symmetry [5]. An "equatorially mirroring" particle (one with infinitesimal bounce amplitude) selects either the northern or southern branch of the equatorial (minimum-B) surface, depending upon the instantaneous value of its bounce phase φ_2 as the particle traverses the singular contour on which $\partial^2 B/\partial s^2=0$ in the plane of symmetry ($\theta=\pi/2$). Lowest-order expansions such as (1.68) apply only to drift shells on which each field line has a single minimum-B point.

II. Pitch-Angle Diffusion

II.1 Violation of an Adiabatic Invariant

Purely adiabatic motion, as described at length in Chapter I, characterizes the dynamical problem in which the phases φ_1, φ_2, and φ_3 are cyclic coordinates. These are the phases canonically conjugate to the fundamental action integrals $J_1 = 2\pi m_0 c|q|^{-1} M$, $J_2 = J$, and $J_3 = (q/c)\Phi$ that identify the three adiabatic invariants M, J, and Φ of charged-particle motion. Strict conservation of M, J, and Φ is only a kinematical ideal that provides the framework for understanding radiation-belt dynamics, and geophysically interesting dynamical phenomena involve violation of one or more of the invariants. Violation of an adiabatic invariant occurs in the presence of forces that vary on so short a spatial or temporal scale that particles having the same three adiabatic invariants (but different phases) respond inequivalently.

Ordinarily this means that violation of the invariant associated with the action integral J_i requires application of a force that varies abruptly on a time scale comparable to the corresponding periodicity of adiabatic motion ($2\pi/\Omega_i$). In some instances, however, spatial symmetries may preserve an invariant even if this condition on the time scale is satisfied. On the other hand, spatial variations of the force field that are abrupt on a length scale comparable to the gyroradius can violate adiabatic invariants, irrespective of the temporal scale.

A variety of geophysical processes can violate the invariants of adiabatic motion. Collisions, for example, act on a scale that is both spatially and temporally abrupt with respect to gyration, and all three of a charged particle's adiabatic invariants can be violated thereby. Electrostatic and electromagnetic plasma cyclotron waves similarly distinguish among particles having different gyration, bounce, and drift phases. Such waves are capable of violating all three adiabatic invariants. Geomagnetic micropulsations typically have frequencies comparable to particle bounce or drift frequencies, and thus can violate J and/or Φ. In many of these examples, the violation of Φ is not severe by comparison with that induced by geomagnetic sudden impulses and other storm- and substorm-associated disturbances of magnetospheric extent. Such disturbances distinguish among particles instantaneously present at different magnetic longitudes (having distinct drift phases

φ_3), but generally average over the phases φ_1 and φ_2, thereby conserving the first two adiabatic invariants.

If the force field responsible for violating the adiabatic invariant associated with an action integral J_i exhibits sufficient spatial and temporal coherence, the distribution of particles initially having in common their values of M, J, and Φ can thereby become organized with respect to the phase φ_i. Then, assuming for simplicity that only one invariant is violated, the associated dispersal of these particles with respect to the conjugate momentum $J_i/2\pi$ can be understood as a consequence of Liouville's theorem (Section I.3)[12]. The dispersal is deterministic in the sense that ΔJ_i, the change in value of J_i, is a function of φ_i; but the dispersal of a particle distribution with respect to J_i appears random if one averages over (or loses sight of) the phase φ_i. In practice, phase mixing always occurs eventually (see Introduction) because any observational instrument has a greater-than-infinitesimal bandwidth with respect to the three invariants. Particles having slightly different values of M, J, and Φ may therefore be counted as being observationally equivalent in the detector. However, since these particles have slightly different values of Ω_i, encompassing a bandwidth $\Delta\Omega_i/2\pi$, their phases φ_i will mix adiabatically on a time scale $\sim 2\pi/\Delta\Omega_i$. Phase memory persists in the distribution as the particles continue to gyrate, bounce, and drift, but this memory is hidden from an observer, to whom the particles appear to be randomly phased (see Introduction).

For this reason, an essentially complete physical description of the earth's radiation environment is provided by specifying the *phase-averaged* particle fluxes (see Introduction) in terms of M, J, Φ, and time. This suppression of the phase variables φ_i introduces an essential component of randomness that permits violation of the adiabatic invariants to be represented by *diffusion* of the particle population with respect to M, J, and/or Φ under most circumstances of interest. After phase averaging, the various elements of the particle distribution, subjected to nonadiabatic forces, usually appear to have walked randomly with respect to the violated invariants. Thus, the ultimate inability to distinguish particle phases by observation is a simplifying virtue.

Since the action variables $J_i/2\pi$ are canonical, the basic form of the diffusion equation for radiation-belt particles is

$$\frac{\partial \overline{f}}{\partial t} = \sum_{ij} \frac{\partial}{\partial J_i}\left[D_{ij}\frac{\partial \overline{f}}{\partial J_j}\right], \qquad (2.01)$$

---

[12]Since the distribution function moves "incompressibly" through phase space in Hamiltonian mechanics [7], a narrowing of the distribution with respect to φ_i implies a broadening with respect to the J_i, and vice versa.

where $\bar{f}(\mathbf{p},\mathbf{r};t)$ is the phase-averaged particle distribution function and D_{ij} is the tensorial diffusion coefficient. For practical purposes, there are only two classes of interaction not describable in terms of (2.01). One class involves change of particle identity, *e.g.*, beta decay, electron attachment, recombination, charge exchange, inelastic capture, nuclear excitation. The other class falls under the general heading of "friction", *e.g.*, the gradual deposition of energy by energetic particles traveling through matter. Where such processes are truly important, as for the inner-zone proton population, it is necessary (and not usually difficult) to add the appropriate source and sink terms to (2.01). Although some of these non-diffusive processes are included below, the primary emphasis of the present work is on that multitude of processes under which (2.01) very adequately describes the behavior of radiation-belt particles.

It is customary in radiation-belt physics to distinguish between pitch-angle diffusion (which violates M or J, and usually both) and radial diffusion (which violates Φ). Although some diffusive processes violate all three invariants, the dichotomous viewpoint is conceptually convenient. As a rule, radial diffusion enables the radiation belts to become populated from an external source (or rearranges particles injected by an internal source), while pitch-angle diffusion causes particle loss to an atmospheric sink. There are exceptions to this rule, but it is often fruitful to think in these terms; hence the distinction between radial diffusion and pitch-angle diffusion. The present chapter is devoted to pitch-angle diffusion, which arises from a variety of mechanisms.

II.2 Collisions

Because radiation-belt particles have such high energies and low densities (see Chapter I), Coulomb collisions between them are completely negligible. Collisions with ionospheric constituents, however, contribute importantly to the ultimate demise of geomagnetically trapped radiation. Energetic particles traveling through matter (including the ionospheric medium) tend to yield their energy to free and bound ambient electrons or to the excitation of atomic nuclei. Moreover, the phenomenon of charge exchange with an ambient atom effectively removes an energetic proton from the radiation-belt population.

As noted above, processes involving *systematic* energy loss to the medium are generally not describable by (2.01). Special terms must be added to account for such *non-diffusive* effects, although the friction mechanism may simultaneously be responsible for diffusion in pitch angle. Because systematic energy loss to the medium can be interpreted as a convective flow of $\bar{f}(\mathbf{p},\mathbf{r};t)$ through adiabatic-invariant space, these special terms have the form of the "divergence" of a non-stochastic "current" in the *Fokker-Planck equation*

$$\frac{\partial \bar{f}}{\partial t} + \sum_i \frac{\partial}{\partial J_i}\left[\left(\frac{dJ_i}{dt}\right)_v \bar{f}\right] = \sum_{ij} \frac{\partial}{\partial J_i}\left[D_{ij}\frac{\partial \bar{f}}{\partial J_j}\right], \tag{2.02}$$

in which the subscript v refers to frictional (non-stochastic) processes. Ordinarily the Fokker-Planck equation is written in the form

$$(\partial \bar{f}/\partial t) = -\sum_i [\partial(D_i' \bar{f})/\partial J_i] + \sum_{ij} [\partial^2(D_{ij}\bar{f})/\partial J_i \partial J_j], \tag{2.03a}$$

where

$$D_i' = (dJ_i/dt)_v + \sum_j (\partial D_{ij}/\partial J_j). \tag{2.03b}$$

The relationship between $\bar{f}$ and the phase-averaged flux $\bar{J}_\alpha$ is given by (1.61), in Section I.7.

Inner-Zone Protons. An important example of non-stochastic "flow" in phase space is the deceleration of inner-zone protons ($M \lesssim 4\,\text{GeV}/$ gauss) by free and bound electrons in the upper ionosphere. Because the rest-mass ratio m_p/m_e is so large, the protons experience no significant range straggling or pitch-angle diffusion (see below) in traversing the medium. In other words, the equatorial pitch angle remains constant while M and J decrease systematically by virtue of energy transfer. The rate of energy transfer is obtained by means of elaborate quantum-mechanical calculations, which yield [37, 38]

$$(m_e v/4\pi q_p^2 q_e^2)(dE/dt)_v = \bar{N}_e[1 - \gamma^{-2} - \ln(\lambda_D m_e v/\hbar)] \tag{2.04}$$
$$+ \sum_i \bar{N}_i Z_i\{1 - \gamma^{-2} - \ln[2m_e c^2(\gamma^2 - 1)/I_i]\},$$

where v is the speed of the proton, and γ is its ratio of relativistic mass to rest mass. The quantities $\bar{N}_e$ and $\bar{N}_i$ are obtained by averaging the densities of free electrons (N_e) and gas molecules (N_i), each of the latter containing Z_i bound electrons, over the proton trajectory (drift shell). Since the ionospheric (or plasmaspheric) Debye length λ_D appears only logarithmically in (2.04), it may be evaluated anywhere on the drift shell (*e.g.*, where $N_e = \bar{N}_e$) without introducing substantial error. The quantity I_i has the significance of a mean excitation energy for the bound electrons; typical values of I_i, along with drift-averaged values of N_e and N_i [38], are given in Table 5 for selected drift shells on which $J = 0$. These shells are identified by the McIlwain parameter L_m (see Section I.5), which equals $(B_0/B_m)^{1/3}$ in the case of particles mirroring at the magnetic equator. The major contribution to each

Table 5. Drift-Averaged Atmospheric Densities, cm^{-3}

L_m	j	$|Z_j|$	I_i, eV	Phase of Solar Cycle		
				Maximum	Averaged	Minimum
1.150	H	1	15	5.36×10^3	7.09×10^3	1.17×10^4
1.186	H	1	15	4.48×10^3	5.90×10^3	9.58×10^3
1.247	H	1	15	3.32×10^3	4.34×10^3	6.93×10^3
1.349	H	1	15	2.23×10^3	2.88×10^3	4.50×10^3
1.500	H	1	15	1.30×10^3	1.65×10^3	2.50×10^3
1.900	H	1	15	5.15×10^2	5.99×10^2	9.22×10^2
2.500	H	1	15	1.95×10^2	2.24×10^2	3.09×10^2
1.150	He	2	41	1.26×10^6	6.17×10^5	1.52×10^5
1.186	He	2	41	5.94×10^5	2.83×10^5	6.51×10^4
1.247	He	2	41	1.74×10^5	7.98×10^4	1.70×10^4
1.349	He	2	41	3.40×10^4	1.48×10^4	2.84×10^3
1.500	He	2	41	3.83×10^3	1.55×10^3	2.61×10^2
1.900	He	2	41	9.64×10^1	2.86×10^1	
2.500	He	2	41	2.25×10^0	7.51×10^{-1}	
1.150	O	8	89	4.83×10^6	2.41×10^6	8.24×10^5
1.186	O	8	89	1.87×10^5	8.41×10^4	1.87×10^4
1.247	O	8	89	9.87×10^2	3.77×10^2	5.92×10^1
1.349	O	8	89		2.82×10^1	
1.150	N_2	14	78	2.45×10^4	1.66×10^4	6.46×10^3
1.186	N_2	14	78	8.06×10^1	4.61×10^1	1.27×10^1
1.150	O_2	16	89	5.57×10^2	3.53×10^2	1.19×10^2
1.150	e	1		1.15×10^5	1.62×10^4	1.61×10^4
1.186	e	1		4.11×10^4	5.34×10^3	7.83×10^3
1.247	e	1		9.32×10^3	2.89×10^3	4.59×10^3
1.349	e	1		3.29×10^3	3.03×10^3	3.24×10^3
1.500	e	1		2.66×10^3	2.65×10^3	2.66×10^3
1.900	e	1		1.69×10^3	1.69×10^3	1.69×10^3
2.500	e	1		7.92×10^2	7.92×10^2	7.92×10^2

$\bar{N}_j$ on these drift shells occurs at the South Atlantic "anomaly" (see Section I.5), where each of the shells attains its perigee altitude under adiabatic motion.

The quantity $E|dE/dt|^{-1}$ is interpreted as an instantaneous e-folding time for the kinetic energy of a proton depositing its energy in the atmosphere. The dependence of this e-folding time on L_m is illustrated in Fig. 14 for protons having selected values of M (given in GeV/gauss) and $J = 0$ [39]. At constant M and J, the "lifetime" against Coulomb deceleration thus peaks at $L_m \approx 1.6$.

In view of the great magnitude of the time scales for proton energy loss (see Fig. 14), it is essential to re-examine assumptions concerning

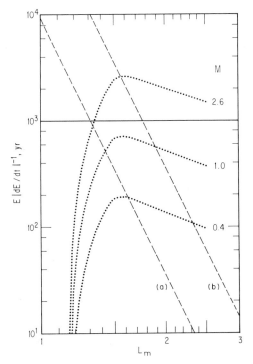

Fig. 14. Effective lifetimes against Coulomb drag (energy loss) for equatorially mirroring inner-zone protons (dotted curves) at selected values of the first invariant M, GeV/gauss. Solid curve shows corresponding time scale $-B_0/2\dot{B}_0$ for energization by present secular variation of geomagnetic dipole moment [39]. Dashed curves show roughly corresponding time scales $L^2/30 D_{LL}$ for energization by inward radial diffusion (see Section III.8), assuming (a) $D_{LL}=10^{-8}L^{10}$ day^{-1} and b) $D_{LL}=10^{-9}L^{10}$ day^{-1}.

the constancy of B_0 in (1.37) when carrying out *theoretical* calculations. In fact, the present value of $\dot{B}_0 (\approx -0.016$ gauss/century) leads to an instantaneous time scale $-(L/\dot{L})=-(B_0/\dot{B}_0)\sim 2000$ yr for the secular contraction of adiabatic drift shells. The conservation of M and J during this secular contraction implies a secular energization of geomagnetically trapped particles. In a contracting dipole field the preservation of $M \equiv p^2 y^2 L^3/2 m_0 B_0$ and $J \equiv 2 L a p Y(y)$ implies $dy/dt=0$ and

$$\frac{1}{E}\frac{dE}{dt}=-\frac{2}{B_0}\frac{dB_0}{dt}\left[\frac{\gamma+1}{2\gamma}\right] \sim \frac{[(\gamma+1)/2\gamma]}{1000\,\mathrm{yr}}. \qquad (2.05)$$

Thus, the equatorial pitch angle remains invariant, and a nonrelativistic proton has its energy increased by a factor of e on a time scale ~ 1000 yr (see Fig. 14). This time scale is comparable to that for energy loss to free and bound electrons, and so it appears that the two processes are mutually competitive for inner-belt protons [39].

Since inner-zone protons are subject, in addition, to radial diffusion over the time scales of interest, a theoretical analysis of the quasi-static profile of the inner belt is deferred to a later chapter (see Section V.7). At this point it is appropriate to discuss only the form of $(dJ_i/dt)_v$ and of its "divergence" with respect to J_i. Since $J_1 = 2\pi m_0 c|q|^{-1} M$ and $J_2 = J$, this "divergence" may be written (for a dipole field) as

$$
\begin{aligned}
\sum_i \frac{\partial}{\partial J_i} \left[\left(\frac{dJ_i}{dt} \right)_v \bar{f} \right] &= \frac{\partial}{\partial M} \left[\left(\frac{dM}{dt} \right)_v \bar{f} \right]_{J,\Phi} + \frac{\partial}{\partial J} \left[\left(\frac{dJ}{dt} \right)_v \bar{f} \right]_{M,\Phi} \\
&= \frac{\partial}{\partial M} \left[\left(\frac{dM}{dt} \right)_v \bar{f} \right]_{\Phi} + \frac{\bar{f}}{2M} \left(\frac{dM}{dt} \right)_v \qquad (2.06) \\
&= \left(\frac{m_0}{2 M B_m^3} \right)^{1/2} \frac{\partial}{\partial M} \left[\gamma^2 v \left(\frac{dE}{dt} \right)_v \bar{f} \right]_{\Phi}
\end{aligned}
$$

in the limit of equatorially mirroring $(J=0)$ protons, for which $(dE/dt)_v$ is given by (2.04)[13]. The unidirectional flux $\bar{J}_{\perp} (\equiv 2 m_0 M B_m \bar{f})$ is considered a function of M and Φ in (2.06) and should be evaluated at the magnetic equator, where $B = B_e$. Of course, both γ and $(dE/dt)_v$ are functions of M and Φ as well; in a dipole field the fact that $B_e = B_0 L^{-3} = (1/8\pi^3 a^6 B_0^2)|\Phi|^3$ implies

$$\gamma^2 = 1 + (2 M B_0/m_0 c^2 L^3) = 1 + (M/4\pi^3 a^6 B_0^2 m_0 c^2)|\Phi|^3 \quad (2.07)$$

for particles having $J=0$. In this case the phase-averaged distribution function that satisfies (2.02) is given by

$$\bar{f} = |\Phi|^{-3} (4\pi^3 a^6 B_0^2/m_0 M) \bar{J}_{\perp}. \qquad (2.08)$$

Coulomb energy loss for ions other than protons can be evaluated from (2.04) if q_p is replaced by the ionic charge and v interpreted as the ionic velocity[14].

Charge Exchange. For protons of much lower energy than those represented in Fig. 14, the main collisional "loss" mechanism is *charge exchange*, whereby a proton absorbs an electron from an ambient atom

---

[13]Note that $v(dE/dt)_v$ depends rather weakly on M, which reduces to $(\gamma^2 - 1) \times (m_0 c^2/2 B_e)$ for $J=0$. It can be shown by Jacobian methods (see below) that (2.06) holds not only for $J=0$, but more generally for any constant value of $K^2 \equiv J^2/8 m_0 M$.

[14]As noted above, the explicit time dependence of B_0 is potentially an important effect for the inner proton belt. The time scales illustrated in Fig. 14 apply only to the present epoch, and not to past or future centuries.

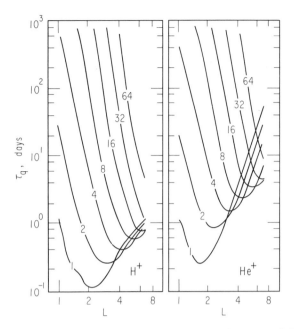

Fig. 15. Charge-exchange lifetimes against neutralization for equatorially mirroring protons (H^+) and helium ions (He^+) at selected values of M/A, MeV/gauss-nucleon [40].

and escapes from the radiation belt as an energetic hydrogen atom. This process is microscopically catastrophic (a "one-shot" interaction), and so, unlike that described by (2.04), it is best characterized by a true lifetime $\tau_q = l_q/v$, where l_q is a mean free path. Typical charge-exchange lifetimes against conversion of H^+ and He^+ ions into H^0 and He^0 atoms by the hydrogen-atom environment are illustrated in Fig. 15 [40] for an appropriate atmospheric model [41]. It is conventional to compare coincident radiation-belt fluxes of distinct ionic species at common values of E/A (kinetic energy per nucleon), where A is the number of nucleons in the ionic nucleus. According to this convention[15], first invariants M for H^+ are directly comparable with first invariants $4M$ for He^+. The particles described in Fig. 15 are nonrelativistic and have vanishing second invariants ($J=0$). The governing equation

---

[15]Because the conventional comparison is between ions having E/A in common at the same point in space, and therefore having v and γ in common within experimental error, the Coulomb "lifetimes" $|(d \ln E/dt)_v|^{-1}$ scale as A_j/q_j^2 according to (2.04), where the subscript j denotes the species of the energetic ion.

in the presence of simple charge-exchange losses has the form

$$\frac{d\bar{f}}{dt} \equiv \frac{\partial \bar{f}}{\partial t} + \sum_i \frac{\partial}{\partial J_i}\left[\left(\frac{dJ_i}{dt}\right)_v \bar{f}\right] - \sum_{ij} \frac{\partial}{\partial J_i}\left[D_{ij}\frac{\partial \bar{f}}{\partial J_j}\right] = -\frac{\bar{f}}{\tau_q}. \quad (2.09)$$

The charge-exchange lifetimes shown in Fig. 15 are deduced from cross sections σ [40] shown in Fig. 16, applied to a model atmosphere [41]

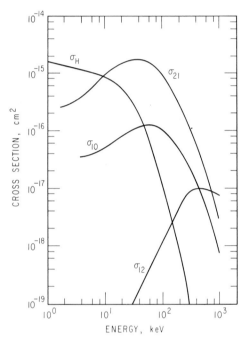

Fig. 16. Ion-energy dependence of charge-exchange cross sections in atomic-hydrogen atmosphere [40].

and model field. Evidently charge transfer is a simple loss process governed by (2.09) only for singly charged ions (with $E \lesssim 400\,\text{keV}$ in the case of He^+). A single cross section σ_H governs conversion of H^+ into H^0, and there are no competing channels open to radiation-belt protons. Three separate cross sections are needed to describe charge transfer in helium: σ_{10} for the neutralization of He^+ ($He^+ \rightarrow He^0$), σ_{12} for the conversion of He^+ into He^{++} (an insignificant reaction for $E \lesssim 400\,\text{keV}$), and σ_{21} for the conversion of He^{++} into He^+ (the largest cross section of the three for $E \lesssim 400\,\text{keV}$). Thus, in the presence of comparable He^+ and alpha-particle (He^{++}) fluxes, it is necessary to introduce coupled transport equations for the phase-averaged distribution functions of He^+ ($\bar{f}_1$) and He^{++} ($\bar{f}_2$), viz.,

$$(d\bar{f}_1/dt) = -(\bar{f}_1/\tau_{10}) - (\bar{f}_1/\tau_{12}) + (\bar{f}_2/\tau_{21}) \qquad (2.10a)$$

$$(d\bar{f}_2/dt) = -(\bar{f}_2/\tau_{21}) + (\bar{f}_1/\tau_{12}). \qquad (2.10b)$$

Except for the possible reconversion of energetic H^0 into H^+ deep within the atmosphere in the course of precipitation (see Section II.7), no such cross coupling occurs in the description of proton or electron radiation belts. This lack of cross coupling is a welcome simplification for these two major constituents of the geomagnetically trapped radiation.

Pitch-Angle Diffusion. In addition to atmospheric deceleration, radiation-belt *electrons* undergo both *pitch-angle diffusion* and *range straggling* to a significant degree. These latter two effects arise because the mass of a radiation-belt electron is equal (apart from relativistic effects) to the mass of the atomic or plasma electrons with which it collides. The result is that deflection (pitch-angle scattering) becomes comparable in importance with energy loss. Moreover, the energy lost in an individual collision strongly depends on the scattering angle, which is a random variable. Thus, atmospheric collisions cause radiation-belt electrons to diffuse not only in pitch-angle cosine (x) but also in energy with respect to the mean value of $(dE/dt)_v$. This latter phenomenon (energy diffusion) is known as *range straggling*, because (in nuclear-physics experimentation) it permits the constituent particles of a monoenergetic beam to traverse statistically varying total path lengths before coming to rest in some material medium. For radiation-belt electrons, range straggling has the effect of smoothing the energy spectrum, which typically arises from a relatively unstructured source spectrum anyway. Thus, range straggling is usually neglected altogether.

In this and other problems for which third-invariant violation is unimportant, the variables E and x (kinetic energy and pitch-angle cosine) are usually more convenient than M and J. The corresponding diffusion matrix

$$\tilde{\mathbb{D}} \equiv \begin{pmatrix} D_{EE} & D_{Ex} \\ D_{xE} & D_{xx} \end{pmatrix} \qquad (2.11)$$

is diagonal because, in individual collisions, ΔE and Δx are statistically uncorrelated; the change in energy is an *even* function of the change in x. Since the ensemble average $\langle \Delta E \Delta x \rangle$ therefore vanishes, so do the off-diagonal components of (2.11).

In general, the transformation of (2.01) and (2.02) from the set of action variables J_i to some set of new variables Q_j requires evaluation of the Jacobian [5] $G(J_i;Q_j) \equiv \det(\partial J_i/\partial Q_j)$. The diffusion operator then has the property that

$$\sum_{ij} \frac{\partial}{\partial J_i}\left[D_{ij}\frac{\partial \bar{f}}{\partial J_j}\right] = \frac{1}{G}\sum_{ij}\frac{\partial}{\partial Q_i}\left[G\tilde{D}_{ij}\frac{\partial \bar{f}}{\partial Q_j}\right]. \qquad (2.12)$$

In a dipole field it is easy to calculate the Jacobian $G(M,J;E,x)$, where $M = p^2 y^2 L^3/2m_0 B_0$ and $J = 2La p\, Y(y)$. According to (1.30), the function $Y(y)$ has the property that $Y(y) - y\, Y'(y) = 2T(y)$. The energy and momentum are related as in (1.60), and $x^2 + y^2 = 1$. The partial derivatives needed for calculating $G(M,J;E,x)$ are

$$(\partial M/\partial E)_x = 2m\, M/p^2 \qquad (2.13\,\text{a})$$

$$(\partial M/\partial x)_E = -2x\, M/y^2 \qquad (2.13\,\text{b})$$

$$(\partial J/\partial E)_x = (2m\, La/p)\, Y(y) \qquad (2.13\,\text{c})$$

$$(\partial J/\partial x)_E = -(2La p\, x/y^2)\, y\, Y'(y), \qquad (2.13\,\text{d})$$

and it follows that

$$G(M,J;E,x) = (4\gamma p L^4 a/B_0)x\, T(y). \qquad (2.14)$$

Moreover, the "divergence" of the non-stochastic "current" introduced in (2.02) may be transformed to read

$$\sum_i \frac{\partial}{\partial J_i}\left[\left(\frac{dJ_i}{dt}\right)_v \bar{f}\right] = \frac{1}{G}\sum_i \frac{\partial}{\partial Q_i}\left[G\left(\frac{dQ_i}{dt}\right)_v \bar{f}\right], \qquad (2.15)$$

and so the Fokker-Planck equation for radiation-belt particles subject only to atmospheric scattering (radial diffusion explicitly ignored) is

$$\frac{\partial \bar{f}}{\partial t} = -\frac{1}{\gamma p}\frac{\partial}{\partial E}\left[\gamma p\left(\frac{dE}{dt}\right)_v \bar{f}\right]_x + \frac{1}{x\,T(y)}\frac{\partial}{\partial x}\left[x\,T(y)\,D_{xx}\frac{\partial \bar{f}}{\partial x}\right]_E$$

$$+ \frac{1}{\gamma p}\frac{\partial}{\partial E}\left[\gamma p\,D_{EE}\frac{\partial \bar{f}}{\partial E}\right]_x. \qquad (2.16)$$

The first term of (2.16) represents a non-stochastic (mean) energy loss to the atmosphere, as described by (2.04). The second term represents pitch-angle diffusion, and the associated transport coefficient is given [42] by

$$D_{xx} = \sum_j \langle (\pi/2\,x^2) v\, N_j [x^2 - 1 + (B_e/B)]$$

$$\times \int_{-1}^{+1} \{2[x^2 - 1 + (B_e/B)](1 - \cos\theta)^2 \qquad (2.17)$$

$$+ (1 - x^2)\sin^2\theta\} (d\sigma_j/d\Omega) d(\cos\theta) \rangle$$

where $d\sigma_j/d\Omega$ is the differential cross section for an energetic electron incident on atmospheric constituent j at scattering angle θ in the "laboratory" frame. Debye shielding is considered in the specification of $d\sigma_j/d\Omega$. The third term of (2.16) represents range straggling (diffusion with respect to energy).

The derivation of (2.17) is straightforward. If an electron initially traveling in the $\hat{\mathbf{z}}$ direction with local pitch angle α relative to $\mathbf{B}$ (which lies locally in the xz plane) is scattered through an angle θ, the resulting change in its value of $\cos\alpha$ is

$$\varDelta \cos\alpha = \cos\alpha(\cos\theta - 1) + \sin\alpha\sin\theta\cos\varphi, \qquad (2.18)$$

where φ is the azimuthal coordinate about the direction of $\hat{\mathbf{z}}$. Since $d\sigma_j/d\Omega$ is independent of φ, the expected value of $2(\varDelta\cos\alpha)^2$ is $2\cos^2\alpha \times (1 - \cos\theta)^2 + \sin^2\alpha\sin^2\theta$. In terms of the equatorial pitch angle $\cos^{-1}x$, one obtains $\sin^2\alpha = 1 - \cos^2\alpha = (1 - x^2)(B/B_e)$. It follows that

$$(\varDelta x)^2 = (B_e/B)^2 \cos^2\alpha (\varDelta\cos\alpha)^2 . \qquad (2.19)$$

Finally, the diffusion coefficient D_{xx} is defined as *half* the rate at which $(\varDelta x)^2$ grows with time.

The factor of one-half that enters the definition of D_{xx} can be understood in terms of a simplified prototype diffusion equation of the form

$$\partial f/\partial t = D_{\xi\xi}(\partial^2 f/\partial\xi^2), \qquad (2.20)$$

which applies in one-dimensional problems for which $D_{\xi\xi}$ is constant with respect to the rectilinear coordinate ξ. The use of (2.20) is chosen over (2.16) for illustrative purposes only because (2.20) is satisfied by the simple unit-normalized Green's function

$$f(\xi, t) = (2\pi a^2)^{-1/2} \exp[-(\xi - \xi_0)^2/2a^2], \qquad (2.21)$$

where a is a function of time and measures the "width" of the distribution in the sense that

$$a^2 = \int_{-\infty}^{+\infty} (\xi - \xi_0)^2 f(\xi, t) d\xi . \qquad (2.22)$$

Direct application of (2.20) implies that

$$D_{\xi\xi} = \frac{d}{dt}\left(\frac{a^2}{2}\right), \qquad (2.23)$$

as indicated. Moreover, the distribution $f(\xi,t)$ given by (2.21) becomes the Dirac function $\delta(\xi-\xi_0)$ in the limit $a=0$. In view of (2.22), the quantity a^2 also represents the net mean-square migration of an individual particle (averaged over the ensemble) from the point $\xi=\xi_0$; the elapsed time of this random migration is $a^2/2D_{\xi\xi}$. This simple illustration epitomizes a general principle that is extremely useful in the calculation of diffusion coefficients from dynamical information. Of course, the metric in (2.16) is not as simple as that in (2.20), and so the Green's function is not easily identified. The basic relationship (between a diffusion coefficient and the ensemble-averaged square of the random migration with respect to a kinematical variable) holds true nevertheless.

Inner-Zone Electrons. Even with the inclusion of Debye shielding, the differential Coulomb cross section $d\sigma_j/d\Omega$ is strongly peaked in favor of forward scattering ($\theta \ll 1$). It follows that the mean value of $(1-\cos\theta)^2$ in (2.17) is *much* smaller than that of $\sin^2\theta$. The distribution of terrestrial atmosphere causes the bulk of the scattering in (2.17) to occur near the

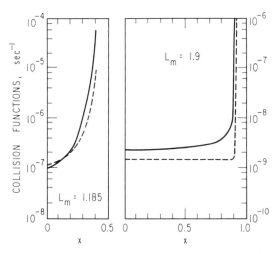

Fig. 17. Magnitudes of $-(v/m_0 c^3)(dE/dt)_v$ (solid curve) and $(2p^3/\gamma m_0^3 c^3 y^2)D_{xx}$ (dashed curve) for inner-zone electrons subjected solely to atmospheric collisions [42].

mirror points, *i.e.*, where $B \approx B_m$. Moreover, the mirror points must not lie too deep in the atmosphere at any longitude if the phase-averaged treatment explicit in (2.01), (2.02), (2.04), (2.16), and (2.17) is to have any meaning. Other methods of analysis, preserving φ_3 as a variable, must be employed unless $2\pi D_{xx} \ll |\Omega_3|$ and $2\pi|(dE/dt)_v| \ll |\Omega_3|E$. Granting these conditions, it is clearly permissible to neglect the term $2[x^2-1 +(B_e/B)](1-\cos^2\theta)$ in (2.17) by comparison with $(1-x^2)\sin^2\theta$. The main energy dependence both of $(dE/dt)_v$ and of $d\sigma_j/d\Omega$ can be factored out, leaving the quantities $-(v/m_0c^3)(dE/dt)_v$ and $(2p^3/\gamma m_0^3 c^3 y^2)D_{xx}$ that are plotted [42] against x in Fig. 17 for selected values of L_m in a model atmosphere. The plotted functions, whose variations with energy are extremely weak, are evaluated here for $E \approx 1.5$ MeV. Both functions have the dimension of frequency.

Figure 17 illustrates a sharp distinction at $L_m = 1.9$ between electrons for which $x \lesssim x_c \approx 0.9$ and those for which $x \gtrsim x_c$. The former are scattered almost negligibly on time scales for which the latter experience virtually immediate absorption by the atmosphere. Equatorial pitch angles for which $|x| > x_c$ are therefore said to constitute an atmospheric *loss cone* in momentum space. In mathematical terms, the coordinates $x = \pm x_c$ represent perfectly absorbing boundaries at which $\bar{f}$ is forced to vanish. Energy loss and pitch-angle diffusion satisfy the conditions $2\pi|(dE/dt)_v| \ll |\Omega_3|E$ and $2\pi D_{xx} \ll |\Omega_3|$ extremely well for $|x| < x_c$ at $E \sim 1$ MeV, thereby justifying the phase-averaged approach. The ultimate sink for inner-zone radiation, however, is quite localized at the South Atlantic "anomaly" (see Section I.5), where drifting particles having $x \approx x_c$ must dip deep into the atmosphere to find their mirror-field intensity B_m.

A rough estimate for the loss-cone angle $\cos^{-1} x_c$ can be obtained by postulating total absorption at altitude $h(\approx 0.02a)$ and displacement of the dipole by $r_0(\approx 0.07a)$ perpendicular to its axis. This eccentricity of the dipole plays an important role in cutting off the inner zone. The indicated parameters predict that

$$1-x_c^2 \approx [(a+h)/La]^3[1+3r_0(La)^{-1/2}(a+h)^{-1/2}]$$
$$\div [4-(3/La)(a+h)-3r_0(La)^{-3/2}(a+h)^{1/2}]^{1/2} \qquad (2.24)$$

as a function of L. This formula yields $x_c \approx 0.940$ at $L = 1.9$ (cf. Fig. 17), $x_c \approx 0.567$ at $L = 1.185$, and $x_c = 0$ at $L \approx 1.085$. Thus, the function $\bar{f}$ should vanish for all pitch angles if $L \lesssim 1.085$; in fact, a true South American anomaly accidentally near the eccentric-dipole "anomaly" raises the lower boundary of the inner zone to $L = 1.10$, approximately. According to Fig. 17, the loss cone is a poorly defined feature at $L_m = 1.185$, and so this simplifying concept is inapplicable there. The loss cone, however, *is* sharply defined over most of the magnetosphere (at least for $L \gtrsim 1.9$, according to Fig. 17), and is known to play an essential role in the dynamics of geomagnetically trapped radiation.

II.3 Wave-Particle Interactions

The atmosphere alone is quite incapable of accounting for the decay rates observed following temporary enhancements of the electron flux beyond $L \approx 1.25$ (see Chapter IV). The situation is indeed extreme at $L \gtrsim 4$, where storm-associated enhancements of the flux at $E \sim 0.5$ MeV characteristically decay by a factor of e on a time scale ~ 5 days [43]; *in situ* deceleration and pitch-angle scattering into the loss cone, if caused solely by collisions with the tenuous atmosphere, would require thousands of years to produce the same amount of decay. The discrepancy is qualitatively similar for outer-belt protons, although the observational data are considerably less extensive than for electrons. It is therefore natural to invoke non-collisional mechanisms for pitch-angle scattering. These mechanisms are classified under the generic term *wave-particle interactions*.

Magnetospheric waves may arise from a variety of sources. Some waves may enter the magnetosphere from the turbulent magnetosheath (see Introduction) [44]. Waves known as *whistlers* originate from lightning discharges in the atmosphere. Whistlers propagate in a plasma wave mode that can also conduct VLF (very low frequency, 3—30 kHz) radio transmissions through the magnetosphere. Man-made (Morse) signals often trigger new VLF emissions in the magnetosphere, as illustrated in Fig. 18. Moreover, plasma instabilities in the whistler (electromagnetic electron-cyclotron) and other wave modes constitute a prodigious magnetospheric source of wave energy. The VLF phenomenon known as *chorus* (see Fig. 18) apparently arises from one such instability. Other plasma instabilities may give rise to waves known as continuous (Pc) and irregular (Pi) *geomagnetic micropulsations*, which are commonly observed on the ground and in space at frequencies from ~ 2 mHz to ~ 1 Hz. A summary [4, 47] of the magnetospherically important frequency classifications is provided in Table 6.

Table 6. Classification of Magnetospheric Signals

Name	Frequency	Name	Period or Rise Time
SHF	3—30 GHz	Pc 1	$2\pi/\omega = 0.2$—5.0 sec
UHF	0.3—3.0 GHz	Pc 2	$2\pi/\omega = 5$—10 sec
VHF	30—300 MHz	Pc 3	$2\pi/\omega = 10$—45 sec
HF	3—30 MHz	Pc 4	$2\pi/\omega = 45$—150 sec
MF	0.3—3.0 MHz	Pc 5	$2\pi/\omega = 150$—600 sec
LF	30—300 kHz		
VLF	3—30 kHz	Pi 1	$\tau_r = 1$—40 sec
ELF	3—3000 Hz	Pi 2	$\tau_r = 40$—150 sec
ULF	$\lesssim 3$ Hz	sc, si	$\tau_r \sim 300$ sec

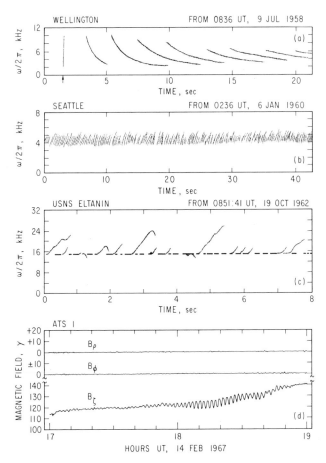

Fig. 18. Examples of magnetospheric wave phenomena observed at $r=a$ [45] and $r=6.6a$ [46]: (a) multiple-hop whistler initiated by nearby lightning stroke (arrow) and reflected between conjugate points along a single magnetospheric path; (b) chorus, characterized by elements of sharply rising frequency; (c) rising and falling VLF emissions triggered by Morse-code transmission from NAA ($\omega/2\pi = 14.7$ kHz, $\Lambda = 56°$) and detected by mobile station at $\Lambda \approx 50°$ in the South Atlantic; (d) coherent Pc-4 micropulsation ($\omega/2\pi \sim 10^{-2}$ Hz) observed at synchronous altitude in the compressional (ζ) component, but absent in the transverse (ρ and φ) components relative to the unperturbed **B** field there.

Not all magnetospheric waves and disturbances can interact effectively with trapped particles; each trapped particle exhibits the three fundamental periodicities of adiabatic motion, and so tends to suppress (filter out) spectral components of applied forces that are distant in frequency from its natural resonances. Thus, trapped particles yield a net diffusive response to forces that have spectral power within a narrow band about some natural resonance frequency in the frame of the particle's adiabatic motion. The width of the passband is determined by the duration of interaction, according to the classical analogue of Heisenberg's uncertainty principle. More specifically, the bandwidth $\Delta\omega$ is equal to $2\pi/\tau$, where τ is the interaction time. The interaction time may be limited by the duration of a wavelike signal or noise burst, by the time required for a particle to traverse a spatially limited region of wave activity, by temporal variation of the wave frequency required by a particle for resonance, or (more generally) by the eventual breakdown of phase coherence between a particle and the Fourier component of the wave spectrum with which it is resonant.

To the extent that the wave spectrum is smooth (structureless) over a bandwidth $\sim 2\pi/\tau$ about a resonance frequency, the interaction is truly resonant in the sense that the "line shape" resembles a Dirac delta function. More generally, the particle accepts a Lorentz-weighted mean of the wave-spectral density over a bandwidth $\Delta(\omega/2\pi)=1/\tau$ about the resonance frequency [16].

In the interest of completeness, Table 6 includes such disturbances as stormtime sudden commencements and magnetic impulses that are only vaguely wavelike in character. More precisely, the "wavelengths" associated with such disturbances are comparable in size to the magnetosphere itself. Since their time scales are so long ($\sim$ minutes), these disturbances violate only the third invariants of radiation-belt particles; such processes are considered in Chapter III. The present chapter is concerned with processes that violate either or both of the first two invariants.

II.4 Bounce Resonance

A force field can violate the second invariant (while preserving the first) through a resonant interaction with the bounce motion of a trapped particle [48]. A force $f_\parallel(s,t)$ that perturbs the bounce motion could typically originate from a compressional (magnetosonic) micropulsation, in which case $f_\parallel = -(M/\gamma)(\partial b_\parallel/\partial s)$, or from an electrostatic wave ($f_\parallel = q_j e_\parallel$). The field perturbations **b** and **e** are understood to project

---

[16]Violation of an adiabatic invariant is the classical analogue of the breakdown of Ehrenfest's theorem in quantum mechanics. This theorem holds that the quantum numbers of a particle, as given by the action integrals of its quasi-periodic motions in the old (Bohr-Sommerfeld) quantum theory, do not change by as much as a unit of $\hbar$ if the applied force field varies only on a sufficiently long time scale. Forces violating this condition lead to diffusion with respect to the classical adiabatic invariants.

nonvanishing components $b_{||} \equiv \mathbf{b} \cdot \hat{\mathbf{B}}$ and $e_{||} \equiv \mathbf{e} \cdot \hat{\mathbf{B}}$ along the unperturbed **B** field. If the normal (to $\hat{\mathbf{B}}$) components of **b** and **e** are confined to the azimuthal ($\hat{\varphi}$) and meridional ($\hat{\varphi} \times \hat{\mathbf{B}}$) directions, respectively, then bounce resonance will not contribute to radial diffusion in an azimuthally symmetric **B** field.

In this case only the second invariant is violated and the governing equation of motion [3] is

$$(dp_{||}/dt) + (M/\gamma)(\partial B/\partial s) = f_{||}(s,t) \tag{2.25}$$

where $p_{||} = \gamma m_0 v_{||} = \gamma m_0 \dot{s}$. Since the unperturbed geomagnetic field **B** is taken to be static, it follows that

$$dw/dt = (p_{||}/m_0)f_{||}(s,t), \tag{2.26}$$

where $w = (p_{||}^2/2m_0) + MB = p^2/2m_0$. The oscillatory force $f_{||}(s,t)$ thus threatens to alter the particle's energy, leaving M and Φ invariant. This is equivalent to the violation of J only, and the relevant Jacobian is

$$G(M,J,\Phi; M,w,\Phi) = (\partial J/\partial w)_{M,\Phi} = 4La(m_0/p)T(y) \tag{2.27}$$

if **B** is given by (1.16), *i. e.*, for a dipole field.

The oscillatory force $f_{||}(s,t)$ is conveniently represented as a superposition of Fourier components applicable to the time interval $0 < t < \tau$. This means that

$$f_{||}(s,t) = \sum_{n=1}^{\infty} f_n \cos(k_{||}s - \omega_n t + \psi_n), \tag{2.28}$$

where $\omega_n = 2\pi n/\tau$, $k_{||}(\mathbf{k} \cdot \hat{\mathbf{B}})$ is the parallel wavenumber corresponding to frequency $\omega_n/2\pi$, and ψ_n is the corresponding phase (ultimately a random variable) of the wavelike Fourier component. Each component contributes $(1/2)f_n^2$ to the mean-square force perturbation $\langle[f_{||}(s,t)]^2\rangle$, and this contribution resides in a frequency interval $\Delta(\omega/2\pi) = \tau^{-1}$. It is therefore appropriate to introduce the spectral density

$$\mathscr{F}_{||}(\omega_n/2\pi) \equiv (\tau/2)f_n^2 \tag{2.29}$$

as an optimal characterization of the force field $f_{||}(s,t)$. Moreover, by virtue of (2.29), each Fourier component acts separately from the others.

The unperturbed bounce motion may be represented

$$s(t) \approx (px/m\Omega_2)\sin(\Omega_2 t + \varphi_2) \tag{2.30}$$

for particles having $x^2 \ll 1$ (see Section I.4). It follows from (2.26)—(2.30) that

$$\Delta w \approx (px/m_0)\,\mathrm{Re}\sum_{n=1}^{\infty}\int_0^{\tau}\cos(\Omega_2 t + \varphi_2) \tag{2.31}$$
$$\times f_n \exp[i(k_n px/m\Omega_2)\sin(\Omega_2 t + \varphi_2) - i\omega_n t + i\psi_n]\,dt\,.$$

If τ is interpreted as a large (ultimately infinite) integral number (N) of bounce periods $2\pi/\Omega_2$, then it follows that $\omega_n = (n/N)\Omega_2$ and

$$\Delta w \approx (p\,x/m_0)\tau \sum_{l=-\infty}^{+\infty} (l/z_l)\,J_l(z_l)\cos(l\,\varphi_2 + \psi_{lN})\,f_{lN} \qquad (2.32)$$

where $z_l \equiv k_{lN}\,px/m\Omega_2$ and J_l denotes the Bessel function of order l.

In the evaluation of $D_{ww} \equiv (1/2\tau)\langle(\Delta w)^2\rangle$, where the angle brackets denote the ensemble average over the phases φ_2 and ψ_{lN}, the cross terms vanish and the result is

$$D_{ww} \approx (p\,x/m_0)^2 \sum_{l=1}^{\infty} (l/z_l)^2\,J_l^2(z_l)\,\mathscr{F}_\|(l\,\Omega_2/2\,\pi)\,. \qquad (2.33)$$

Since $(\partial w/\partial x)_{M,L} = 2\,M\,B_0\,x/L^3 y^4$, it is logical to define a diffusion coefficient

$$D_{xx} = [(\partial w/\partial x)_{M,L}]^{-2}\,D_{ww}$$
$$\approx (L^3 y^6/2m_0\,M\,B_0) \sum_{l=1}^{\infty} (l/z_l)^2\,J_l^2(z_l)\,\mathscr{F}_\|(l\,\Omega_2/2\,\pi)\,. \qquad (2.34)$$

The corresponding diffusion equation

$$\frac{\partial \bar{f}}{\partial t} = \frac{y^3}{x\,T(y)}\,\frac{\partial}{\partial x}\left[\frac{x\,T(y)}{y^3}\,D_{xx}\,\frac{\partial \bar{f}}{\partial x}\right]_{M,L} \qquad (2.35)$$

is constructed from the canonical formalism by inserting the Jacobian $G(M,J;M,x) = (\partial J/\partial w)_{M,L}(\partial w/\partial x)_{M,L} = (8\,am_0\,M\,B_0/L^2)(x/py^4)\,T(y)$ in (2.12). The diffusion coefficient D_{xx} displays a strongly inverse dependence on $x^2(\equiv 1-y^2)$, if only because of the factor y^6. In addition, the Bessel functions act to suppress D_{xx} for particles that mirror beyond a "wavelength" from the equator, i.e., for $z_l \gtrsim l$. This justifies the approximation, inherent in (2.30), that bounce resonance acts principally on particles having $x^2 \ll 1$.

If the origin for $f_\|(s,t)$ is a spectrum of compressional (magnetosonic) micropulsations, then (in a cold plasma) the relevant dispersion relation is $\omega = c_A k$, where c_A is the Alfvén speed [49]. In this case it is found that $z_l = l(k_\|/k)(px/mc_A)$ and that $b_\| = (k_\perp/k)b$. Since $f_\|(s,t) = -(M/\gamma)(\partial b_\|/\partial s)$, the spectral density $\mathscr{F}_\|(\omega/2\pi)$ has the property that

$$\mathscr{F}_\|(\omega/2\pi) = (M/\gamma)^2\,k_\|^2\,\mathscr{B}_\|(\omega/2\pi) \qquad (2.36)$$

where $\mathscr{B}_\|(\omega/2\pi)$ is the spectral density of the magnetic-field perturbation $b_\|$. It is tacitly understood that (2.34) represents an average over the direction of propagation $\hat{\mathbf{k}}$, weighted according to the relative contribution of each $\hat{\mathbf{k}}$ to $\mathscr{B}_\|(\omega/2\pi)$.

Alternatively, if the oscillating force $f_{||}(s,t)$ arises from a spectrum of electrostatic waves, then the force is given by $f_{||}(s,t) = -q(\partial \varphi / \partial s)$, where $\varphi(s,t)$ is the oscillating electrostatic potential. For this situation it follows that $z_l = l(px/m)(k_{||}/\omega)$, where $\omega = l\Omega_2$, and that

$$\mathscr{F}_{||}(\omega/2\pi) = q^2 k_{||}^2 \, \mathscr{V}(\omega/2\pi), \tag{2.37}$$

where $\mathscr{V}(\omega/2\pi)$ is the spectral density of $\varphi(s,t)$. The values of $k_{||}$ appearing in (2.36) and (2.37) are related to ω by the dispersion relation appropriate to the wave mode in question.

Pitch-angle diffusion by bounce resonance has the distinctive property that particles having $x^2 \ll 1$ are much more strongly affected than those for which $x^2 \sim 1$. This means that bounce resonance may diffuse the mirror points of trapped particles to perhaps $\sim 20°$ latitude from the magnetic equator, whereupon some complementary process, acting preferentially upon particles for which $x^2 \sim 1$, must complete the task of diffusing their equatorial pitch angles into the loss cone [43, 48].

II.5 Cyclotron Resonance

Particles that do not mirror at the equator often satisfy a resonance condition of the form

$$\omega - k_{||} v_{||} - l\Omega_1 = 0; \quad l = \pm 1, \pm 2, \ldots \tag{2.38}$$

with electromagnetic or electrostatic plasma cyclotron waves. This condition is known as Doppler-shifted local *cyclotron resonance*. If the wave frequency $\omega/2\pi$ is held constant, then $k_{||}$ varies with position along the field line. Both $v_{||}$ and Ω_1 vary with the position of a particle's guiding center in the course of its bounce motion. This is the sense in which cyclotron resonance is a *local* phenomenon; the conditions that satisfy (2.38) do not persist over the entire bounce path. Cyclotron resonances therefore have an intrinsic breadth $\Delta\omega = 2\pi/\tau$, where the optimal interaction time τ is estimated from the expression

$$2\pi/\tau = \text{Max}\left[\dot\omega\tau, (\ddot\omega/8)\tau^2\right] = \text{Max}\left[(2\pi\dot\omega)^{1/2}, (\pi^2\ddot\omega/2)^{1/3}\right]. \tag{2.39}$$

The symbols $\dot\omega$ and $\ddot\omega$ represent time derivatives of the value of ω required for resonance as the particle proceeds to execute its adiabatic bounce motion. Since $\dot\omega = 0$ at the equator and at the mirror points, the optimal interaction time is then given by $\tau^3 = 16\pi/\ddot\omega$. Very roughly speaking, this means that $\Delta\omega \sim \varepsilon^{2/3}\Omega_1$. Local resonance, other than near points where $\dot\omega = 0$, has an optimal interaction time given by $\tau^2 = 2\pi/\dot\omega$ and a minimum bandwidth $\Delta\omega \sim \varepsilon^{1/2}\Omega_1$. The case $\dot\omega = 0$ appears somewhat the more favorable for sharp resonance, *i.e.*, for minimizing $\Delta\omega$.

Electrostatic Waves. Both of the above cases find τ to be substantially smaller than the bounce period (since $\varepsilon \ll 1$), and so the analytical problem can be treated in terms of a locally uniform **B** field. Diffusion of the local pitch angle α is equivalent to diffusion in x by virtue of (2.19). In a uniform **B** field containing only particles and electrostatic waves, the equation of motion for each Fourier component is

$$\dot{\mathbf{p}} + \Omega_1(\mathbf{p} \times \hat{\mathbf{B}}) = q e_k \hat{\mathbf{k}} \exp(i\mathbf{k} \cdot \mathbf{r} - i\omega t + i\psi_k) \tag{2.40}$$

where $\Omega_1 = -qB/mc$. The component D_{xx} of the diffusion tensor [see (2.11)] is obtained from the local diffusion coefficient for $\cos\alpha \equiv p_{\parallel}/p$, where $p_{\parallel} = \mathbf{p} \cdot \hat{\mathbf{B}}$. There is no loss of generality in taking $k_x = 0$ and $k_y = k_{\perp}$ in the equation $\int \cos\alpha \, dt$

$$\Delta(\cos\alpha)_k = \mathrm{Re} \int_0^\tau (q/p^3 k) e_k [p_{\perp}^2 k_{\parallel} - k_x p_{\parallel} p_x - k_y p_{\parallel} p_y]$$
$$\times \exp(ik_x x + ik_y y + ik_{\parallel} z - i\omega t + i\psi_k) dt, \tag{2.41}$$

which follows from (2.40). Insertion of the unperturbed orbit $[x = (p_{\perp}c/qB)\cos(\Omega_1 t + \varphi_1); \quad y = (p_{\perp}c/qB)\sin(\Omega_1 t + \varphi_1); \quad z = v_{\parallel}t]$ then yields

$$\Delta(\cos\alpha)_k = \sum_{l=-\infty}^{+\infty} (q/p^3 k) e_k J_l(k_{\perp} p_{\perp} c/qB)[p_{\perp}^2 k_{\parallel}$$
$$+ l(qB/c)p_{\parallel}]\tau \cos(l\varphi_1 + \psi_k) \tag{2.42}$$

to the required first order of accuracy in e_k, upon application of (2.38). Evaluation of the phase averages over φ_1 and ψ_k finally implies

$$D_{xx} = \langle (B_e/B)(y/x)^2 (q^2/2p^2)\cos^2\alpha$$
$$\times \sum_{l=-\infty}^{+\infty} J_l^2(k_{\perp} v_{\perp}/\Omega_1) \mathscr{V}(\omega/2\pi)[k_{\parallel}\sin\alpha - l(\Omega_1/v_{\perp})\cos\alpha]^2 \rangle, \tag{2.43}$$

where the angle brackets denote a bounce average. The spectral density $\mathscr{V}(\omega/2\pi)$ is evaluated at the resonant ω given by (2.38) for each l. The weighted average over the various directions of $\hat{\mathbf{k}}$ present in the spectrum is tacitly understood in (2.43), as in (2.34).

The three remaining components of (2.11) do not vanish, but D_{xx} is the component of primary significance in the analysis of pitch-angle diffusion. Components such as D_{EE} and D_{xE} enable the waves to exchange energy with the particles. The direction of this energy exchange depends upon the form of the particle distribution function $\bar{f}(\mathbf{p}, \mathbf{r})$, and leads accordingly either to amplification or attenuation of waves having a given value of **k** in the mode of interest. Thus, the free energy present in a non-Maxwellian particle distribution can be extracted by the available wave modes under certain conditions. This can lead to the spontaneous generation of plasma waves in the magnetosphere.

Electromagnetic Waves. The electromagnetic cyclotron modes of a plasma are of great geophysical interest in the context of spontaneous wave generation. For propagation along the magnetic field, these modes are circularly polarized, such that the magnetic-field perturbation is given by

$$b_x = b_\perp \cos(k_\| z - \omega t + \psi_k) \tag{2.44a}$$

$$b_y = \pm b_\perp \sin(k_\| z - \omega t + \psi_k). \tag{2.44b}$$

Particles in a locally uniform unperturbed **B** field follow the equation of motion

$$\dot{\mathbf{p}} + \Omega_1 \mathbf{p} \times \hat{\mathbf{B}} = q\mathbf{e} + (q/mc)\mathbf{p} \times \mathbf{b} \tag{2.45}$$

when subjected to (2.44). The induced electric-field perturbation **e** is given by $n\mathbf{e} = -\hat{\mathbf{B}} \times \mathbf{b}$, where $n(\equiv ck_\|/\omega)$ is the refractive index.

If τ is the interaction time, then the first-order change in $\cos\alpha (\equiv p_\|/p)$ is given by

$$\Delta(\cos\alpha)_k = (q/np^2)[n(p/mc) - \cos\alpha] \int_0^\tau (p_x b_y - p_y b_x) dt$$

$$= (q/np)[n(p/mc) - \cos\alpha]\tau b_\perp \sin\alpha \tag{2.46}$$

$$\times \cos(\psi_k \mp \varphi_1); \quad \omega - k_\| v_\| = \mp \Omega_1.$$

The upper sign in (2.44b) therefore leads to cyclotron resonance for $\omega - k_\| v_\| + \Omega_1 = 0$; the opposite polarization implies the resonance condition $\omega - k_\| v_\| - \Omega_1 = 0$. The required phase averages yield

$$D_{xx} = \langle (B_e/B)(y/x)^2 \cos^2\alpha (q^2/2n^2 p^2)$$

$$\times [n(p/mc) - \cos\alpha]^2 \mathscr{B}_\perp(\omega/2\pi) \rangle \tag{2.47}$$

upon application of (2.19), (1.22), and (1.05). The term np/mc dominates $\cos\alpha$ if $\omega \ll |\Omega_1|$, in which case the diffusion is approximately elastic $(\dot{p}_\|^2 \approx \dot{p}_\perp^2 \gg \dot{p}^2)$.

It is instructive to recast the equation of motion as

$$p_\| \dot{p}_\| = q(p_\|/mc)(p_x b_y - p_y b_x) \tag{2.48a}$$

$$p_\perp \dot{p}_\perp = (q/n)[1 - (np_\|/mc)](p_x b_y - p_y b_x). \tag{2.48b}$$

In terms of the phase velocity $v_p = \omega/k_\|$, it follows that

$$d(p^2)/dt = 2(mc/n)\dot{p}_\| = (v_p/v_\|)[d(p_\|^2)/dt] \tag{2.49a}$$

or

$$p_\perp^2 + 2\int (p_\| - m v_p) d p_\| = \text{constant} . \qquad (2.49\,\text{b})$$

It is understood here that $\omega > 0$, that $\Omega_1 < 0$ for $q > 0$, and that $\Omega_1 > 0$ for $q < 0$. (This convention is not universally accepted.) The case $k_\| v_\| < 0$ corresponds to the normal Doppler shift (particle and wave traveling in opposite directions). It allows ions to resonate with ion-cyclotron waves (sometimes called LH for their left-handed polarization relative to **B**) and electrons to resonate with electron-cyclotron waves (otherwise known as RH or whistler-mode waves). According to (2.49), the wave gains energy from any resonant particle whose pitch angle thereby decreases. In other words, the conversion of $p_\perp^2$ into $p_\|^2$ is accompanied by the loss of particle energy to the wave, since $d(p^2)/dt < 0$ (see Fig. 19 [54]).

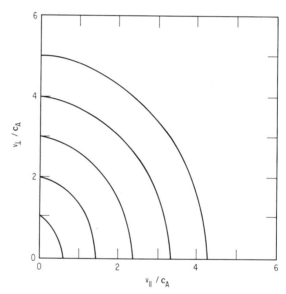

Fig. 19. Velocity-space trajectories of nonrelativistic protons resonant with electromagnetic proton-cyclotron waves propagating parallel to a uniform magnetic field [54].

Since pitch-angle diffusion tends to drive the distribution $\bar{f}$ toward pitch-angle isotropy, the anisotropy caused by the presence of a loss cone (see Section II.2) at $\alpha \approx 0$ represents a source of free energy for the amplification of ion- and electron-cyclotron waves. This is true because the pitch-angle diffusion produces a net diffusive flow of pitch angles into the loss cone, thereby converting $p_\perp^2$ into $p_\parallel^2$ for the typical resonant particle[17].

For $k_\perp = 0$ the electromagnetic cyclotron modes in a cold plasma satisfy dispersion relations of the form

$$(c k_\parallel/\omega)^2 = 1 - \sum_j [\omega_j^2/\omega(\omega \pm \Omega_j)] \tag{2.50}$$

where the subscript j denotes particle species. The ion or electron plasma frequency $\omega_j/2\pi$ is given by $\omega_j^2 = 4\pi N_j q_j^2/m_j$, where N_j is the particle number density, q_j is the particle charge, and m_j is the particle rest mass. The nonrelativistic gyrofrequency $\Omega_j/2\pi$ is given by the formula $\Omega_j = -q_j B/m_j c$, so as to agree in sign with the definition of Ω_1 (see Section I.1; many authors define $\Omega_j \equiv +q_j B/m_j c$). The choice of sign in (2.50) depends upon the sense of circular polarization relative to **B**, as in (2.44). The upper sign corresponds to an ion-cyclotron mode, and the lower to an electron-cyclotron mode. In a two-component plasma ($j = e$ for electrons and $j = i$ for ions) it is customary to simplify (2.50) thus:

$$(c k_\parallel/\omega)^2 \approx \omega_e^2/\omega(\Omega_e - \omega); \qquad \omega \gg |\Omega_i| \tag{2.51a}$$

$$(\omega/k_\parallel)^2 \approx c_A^2(1 - |\omega/\Omega_i|); \qquad \omega < |\Omega_i|. \tag{2.51b}$$

The Alfvén speed $c_A \equiv (B^2/4\pi m_i N_i)^{1/2}$ is presumed to be much smaller than the speed of light c.

Various Types of Cyclotron Resonance. Resonance via the "normal" Doppler shift occurs for $k_\parallel v_\parallel = \omega - |\Omega_j/\gamma| < 0$. For electrons thus resonating with (2.51a), the required whistler-mode wave frequency is related to the particle kinetic energy $(\gamma - 1)m_e c^2$ and local pitch angle α by

$$(c/c_A)^2 \cos^2 \alpha = \frac{[1 - \gamma(\omega/\Omega_e)]^2 [1 - (\omega/\Omega_e)]}{(m_e/m_i)(\omega/\Omega_e)(\gamma^2 - 1)}. \tag{2.52a}$$

For ions of velocity **v** resonating analogously with (2.51b), the corresponding relationship is

$$\cos^2 \alpha = \frac{(1 - |\gamma \omega/\Omega_i|)^2 (1 - |\omega/\Omega_i|)}{(v/c_A)^2 (\omega/\Omega_i)^2 \gamma^2}. \tag{2.52b}$$

---

[17]A word of caution is in order here. The path along which particles diffuse in momentum space $(p_\parallel, p_\perp)$ is not a path of constant p. Thus, wave growth requires essentially that $-\nabla_p \bar{f}$ point toward increasing $p_\parallel$ (decreasing $p_\perp$) along the path of diffusion. This condition is somewhat more stringent than the minimal requirement of pitch-angle anisotropy [i.e., $(\partial \bar{f}/\partial y)_E > 0$] at constant energy. A more quantitative stability analysis is given below (see Section II.6).

Figure 20 indicates the normalized frequency $|\omega/\Omega_j|$ required for resonance according to (2.52). The two cases of interest, electrons ($j = e$) resonating with (2.51 a) and nonrelativistic ions ($j = i$) with (2.51 b), are plotted separately.

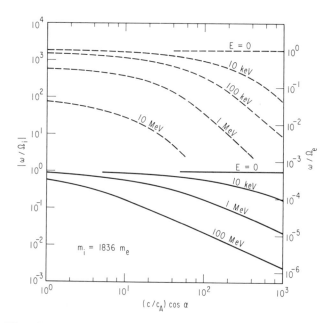

Fig. 20. Wave frequencies required for electromagnetic cyclotron resonance with protons (solid curves) and electrons (dashed curves) via the "normal" Doppler shift, assuming **k** parallel to **B**. Termination of electron-wave contours at $\omega \sim 2|\Omega_i|$ is dictated by use of (2.51 a).

For a given particle, the minimum wave frequency $\omega/2\pi$ required for resonance is that required at the equator, where $v_{||}$ attains its maximum value and $|\Omega_j|$ its minimum along the bounce path. The distribution of particle density (hence, refractive index) along the field line cannot overcome this tendency unless B^2/N_j *decreases* with increasing B. Such a distribution of N_j occurs only in low-altitude regions where Coulomb collisions are already more important than wave-particle interactions in radiation-belt dynamics.

To the extent that $\mathcal{B}_\perp(\omega/2\pi)$ tends to fall with rising frequency in the resonant region, the "normal" Doppler-shifted cyclotron resonance acts preferentially upon particles that mirror away from the equator (see Fig. 20). This mechanism for pitch-angle diffusion is thus complementary to bounce resonance, which acts preferentially upon particles having

$x^2 \ll 1$, in disposing of the particles that populate the earth's radiation belts [43].

A second form of cyclotron resonance involves the "overtaking" or "anomalous" Doppler shift $(k_\parallel v_\parallel = \omega + |\Omega_j/\gamma| > 0)$, whereby the particle sees a wave with its sense of circular polarization apparently reversed [50]. The "anomalous" Doppler shift enables ions to resonate with (2.51a) if

$$\cos^2 \alpha \approx (c/nv)^2 \approx (m_e/m_i)(\omega/\Omega_e)[1 - (\omega/\Omega_e)](c_A/v)^2, \qquad (2.53\,a)$$

where $n(\equiv ck_\parallel/\omega)$ is the refractive index. Moreover, electrons can thus resonate with (2.51 b)

$$(c/c_A)^2 \cos^2 \alpha \approx (\Omega_e/\omega)^2(\gamma^2 - 1)^{-1}(1 - |\omega/\Omega_i|)^{-1}. \qquad (2.53\,b)$$

This last interaction is believed to be responsible for the precipitation of relativistic electrons $(\gamma \gtrsim 4)$ during the main and early recovery phases of a magnetic storm [51]. Figure 21 indicates the normalized wave frequencies $|\omega/\Omega_j|$ with which protons and electrons can resonate via the "anomalous" Doppler shift.

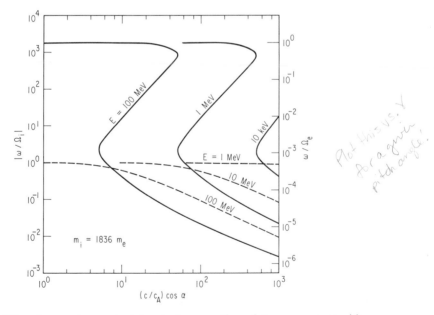

Fig. 21. Wave frequencies required for electromagnetic cyclotron resonance with protons (solid curves) and electrons (dashed curves) via the "anomalous" Doppler shift, assuming **k** parallel to **B**. Extension of proton contours to $\omega \lesssim |\Omega_i|$ is effected by using (2.50) in place of (2.51a).

For the pitch-angle anisotropy characteristic of a loss-cone distribution $\bar{f}$, the "anomalous" Doppler shift leads to a resonant pitch-angle diffusion that extracts energy from the wave spectrum. Moreover, the restriction that $k_\perp = 0$ makes (2.44)—(2.53) a somewhat oversimplified description of geophysical reality. The acceptance of $k_\perp \neq 0$ introduces cyclotron-harmonic resonances $(l \neq \pm 1)$ accompanied by squared Bessel functions $J_l^2(k_\perp v_\perp / \Omega_1)$ in D_{xx}. Often these resonances also extract energy from the wave spectrum. Such wave-absorbing resonances are called *parasitic* [53], since they detract from the wave-amplifying properties of (2.52) in the presence of a loss-cone distribution. Ordinarily, however, the parasitic resonances account for only a fraction of the energy transfer between the particle distribution and wave spectrum, since they tend to be associated (at a given ω) with the more sparsely populated (high-energy) portion of $\bar{f}$ than the primary resonances described by (2.52). Thus, the wave-amplifying properties of $\bar{f}$ remain largely intact.

II.6 Limit on Trapped Flux

Plasma instabilities driven by radiation-belt particles are of special importance in that they can sometimes enforce an upper limit on the particle flux trapped by the geomagnetic field [52]. An instability analysis of the electromagnetic cyclotron modes is best formulated in terms of the plasma dielectric tensor derived from (1.12). The required operations can be simplified by taking $k_\perp = 0$, $e_{||} = 0$, and

$$e_x = \pm i e_\perp \exp(i k_{||} z - i\omega t + i\psi_k) \tag{2.54a}$$

$$e_y = e_\perp \exp(i k_{||} z - i\omega t + i\psi_k). \tag{2.54b}$$

The real part of (2.54) agrees with (2.44) if $e_\perp = -(\omega/c k_{||})b_\perp$. The Maxwell equations for this Fourier component of $\mathbf{e}$ and $\mathbf{b}$ thus read

$$\mathbf{k} \cdot \mathbf{e} = \mathbf{k} \cdot \mathbf{b} = 0, \tag{2.55a}$$

$$c\mathbf{k} \times \mathbf{e} = \omega \mathbf{b}, \tag{2.55b}$$

$$c\mathbf{k} \times \mathbf{b} = -4\pi i \mathbf{J} - \omega \mathbf{e}, \tag{2.55c}$$

where $\mathbf{J}$ is the current-density perturbation. The vanishing divergence of $\mathbf{e}$ is characteristic of the electromagnetic cyclotron wave modes at $k_\perp = 0$, since the perturbation of net charge density vanishes.

Dispersion Relation. It follows for these wave modes that

$$(c^2 k^2 - \omega^2)\mathbf{e} = 4\pi i \omega \mathbf{J}. \tag{2.56}$$

If the distribution function $f_j(\mathbf{p},\mathbf{r};t)$ for species j is decomposed into a phase-averaged part $\bar{f}_j(\mathbf{p},\mathbf{r})$ plus an oscillatory part $\tilde{f}_j(\mathbf{p},\mathbf{r};t)$, the Vlasov equation (1.12) can be written in the linearized form

$$
\begin{aligned}
-i(\omega &- k_\parallel v_\parallel)\tilde{f}_j - (\Omega_j/\gamma_j)\mathbf{p}\times\hat{\mathbf{B}}\cdot\mathbf{V}_p\tilde{f}_j \\
&= -q_j\mathbf{e}\cdot\mathbf{V}_p\bar{f}_j - q_j(k_\parallel/\omega)\mathbf{v}\times(\hat{\mathbf{k}}\times\mathbf{e})\cdot\mathbf{V}_p\bar{f}_j \\
&= -(q_j/\omega)(\omega-k_\parallel v_\parallel)\mathbf{e}\cdot\mathbf{V}_p\bar{f}_j - q_j(k_\parallel/\omega)(\mathbf{e}\cdot\mathbf{v})(\partial\bar{f}_j/\partial p_\parallel),
\end{aligned}
\tag{2.57}
$$

where γ_j is the ratio of relativistic mass to rest mass. By transforming to the variables $p_\perp$ and φ_1, such that $p_x = -p_\perp\sin\varphi_1$ and $p_y = p_\perp\cos\varphi_1$ during unperturbed gyration, it can be shown that

$$
\begin{aligned}
(\Omega_j/\gamma_j)(\partial\tilde{f}_j/\partial\varphi_1) &- i(\omega-k_\parallel v_\parallel)\tilde{f}_j \\
&= -[(q_j/\omega p_\perp)(\omega-k_\parallel v_\parallel)(\partial\bar{f}_j/\partial p_\perp) \\
&\quad + (q_j k_\parallel/\gamma_j m_j\omega)(\partial\bar{f}_j/\partial p_\parallel)](e_x p_x + e_y p_y) \\
&\quad + (q_j/\omega p_\perp^2)(\omega-k_\parallel v_\parallel)(\partial\bar{f}_j/\partial\varphi_1)(e_x p_y - e_y p_x).
\end{aligned}
\tag{2.58}
$$

The final line of (2.58) vanishes because $\bar{f}_j$ is phase-averaged, and therefore independent of φ_1.

It is assumed that ω has a small imaginary part that describes the damping ($\operatorname{Im}\omega < 0$) or growth ($\operatorname{Im}\omega > 0$) of the wave. It is proper to view (2.58) as an ordinary differential equation for $\tilde{f}_j(\varphi_1)$, subject to the "boundary" condition that $\tilde{f}_j(-\infty) = 0$ for $\operatorname{Im}\omega > 0$ and $\tilde{f}_j(+\infty) = 0$ for $\operatorname{Im}\omega < 0$. Thus, the solution of (2.58) may be written

$$
\begin{aligned}
\tilde{f}_j(\varphi_1) = (q_j/2\omega)&[(\omega-k_\parallel v_\parallel)(\partial\bar{f}_j/\partial p_\perp) + k_\parallel v_\perp(\partial\bar{f}_j/\partial p_\parallel) \\
&\times\left[\frac{(e_x - ie_y)\exp(i\varphi_1)}{\omega-k_\parallel v_\parallel - (\Omega_j/\gamma_j)} - \frac{(e_x + ie_y)\exp(-i\varphi_1)}{\omega-k_\parallel v_\parallel + (\Omega_j/\gamma_j)}\right].
\end{aligned}
\tag{2.59}
$$

The electric current-density perturbation $\mathbf{J}$ can thus be written

$$
\begin{aligned}
\mathbf{J} &= \sum_j q_j\int v_\perp(\hat{\mathbf{y}}\cos\varphi_1 - \hat{\mathbf{x}}\sin\varphi_1)\tilde{f}_j(\varphi_1)\,d^3p \\
&= \sum_j (q_j^2/2i\omega)\,\mathbf{e}\int\frac{[(\omega-k_\parallel v_\parallel)(\partial\bar{f}_j/\partial p_\perp) + k_\parallel v_\perp(\partial\bar{f}_j/\partial p_\parallel)]v_\perp\,d^3p}{\omega-k_\parallel v_\parallel \pm(\Omega_j/\gamma_j)}
\end{aligned}
\tag{2.60}
$$

where $d^3p = p_\perp\,dp_\perp\,dp_\parallel\,d\varphi_1$, and where the choice of sign ($\pm$) corresponds to the choice of polarization ($e_x = \pm ie_y$) in (2.54). In terms of the unit-normalized distribution function $F_j = N_j^{-1}\bar{f}_j$, the dispersion relation deduced from (2.56) for $k_\perp = 0$ is therefore

$$
c^2 k^2 = \omega^2 + \pi\sum_j\omega_j^2 I_j,
\tag{2.61 a}
$$

where

$$\int\limits_{-\infty}^{+\infty} \int\limits_{0}^{\infty} \frac{[(\omega - k_\parallel v_\parallel)(\partial F_j/\partial p_\perp) + k_\parallel v_\perp (\partial F_j/\partial p_\parallel)] p_\perp^2 \, dp_\perp \, dp_\parallel}{\gamma_j(\omega - k_\parallel v_\parallel) \pm \Omega_j}. \qquad (2.61\,\text{b})$$

In the rest frame of a cold plasma whose various components j exhibit no relative streaming along **B**, the integration of each I_j by parts allows (2.50) to be recovered from (2.61).

Growth Rate. If the cold plasma is augmented by a comparatively *small* density of hot plasma or of radiation-belt particles, then (2.50) remains approximately valid for relating $k_\parallel$ to the real part of ω. The growth rate Im ω follows directly from (2.61). If $\nabla_p \bar{f}$ is free of substantial variation in the velocity interval extending $\sim |(3 \operatorname{Im}\omega/k_\parallel)|$ to either side of the resonant velocity $v_r \equiv (\omega/k_\parallel) \pm (\Omega_j/\gamma_j k_\parallel)$, then the integral over $p_\parallel$ can be simplified by means of the formula

$$[\gamma_j(\omega - k_\parallel v_\parallel) \pm \Omega_j]^{-1} \approx \mathrm{P}\{[\gamma_j(\omega - k_\parallel v_\parallel) \pm \Omega_j]^{-1}\}$$
$$- i\pi m_j |k_\parallel|^{-1} \delta(p_\parallel - \gamma_j m_j v_r), \qquad (2.62)$$

where P denotes the Cauchy principal value and δ the Dirac function. It follows that

$$c^2 k^2 \approx \omega^2 - \sum_j [\omega_j^2 \, \omega/(\omega \pm \Omega_j)]$$
$$- 4\pi^3 i |k_\parallel|^{-1} \sum_j q_j^2 \int\limits_0^\infty [\mp (\Omega_j/\gamma_j)(\partial \bar{f}_j/\partial p_\perp) \qquad (2.63)$$
$$+ k_\parallel v_\perp (\partial \bar{f}_j/\partial p_\parallel)] p_\perp^2 \, dp_\perp,$$

where the integral follows the path $p_\parallel = \gamma_j m_j(\omega/k_\parallel) \mp (q_j B/ck_\parallel)$.

It is inconvenient to evaluate (2.63) in its full relativistic generality. Since the resonant protons and electrons that tend to amplify magnetospheric cyclotron waves are typically nonrelativistic, it is reasonable to adopt the nonrelativistic ($\gamma_j = 1$) approximation for evaluating (2.63), in which case the integral follows a path of constant $p_\parallel$. In this approximation, a particle distribution whose energy spectrum follows a power law (E^{-l}) and whose pitch-angle distribution approximates $\sin^{2s}\alpha$ over the unit sphere (l and s are not necessarily integers) may be represented as

$$\bar{f}_j(\mathbf{p}, \mathbf{r}) = (p_j/p)^{2l}(p_\perp/p)^{2s} p^{-2} J_\perp(p_j^2/2m_j), \qquad (2.64)$$

where p_j is the scalar momentum that corresponds to a kinetic energy $p_j^2/2m_j$. The form of $\bar{f}_j$ given by (2.64) is observationally realistic, as well as algebraically convenient. It leads (2.63) to predict a growth rate Im ω, given in lowest order by taking the imaginary part of (2.63). The result is

$$\mathrm{Im}\,\omega \approx -4\pi^3 |k_\|{}|^{-1} \sum_j q_j^2 (p_j/m_j\,v_r)^{2l}\,B(s+1,l)[\omega + s(\omega \pm \Omega_j)]\,J_\perp(p_j^2/2m_j)$$
$$\div \{2\omega \mp \sum_j [\omega_j^2\,\Omega_j/(\omega \pm \Omega_j)^2]\}, \qquad (2.65)$$

where $B(s+1,l) \equiv \Gamma(s+1)\Gamma(l)/\Gamma(s+l+1)$ is the beta function. The denominator of (2.65) can be expressed as $2\omega(c^2/v_p v_g)$, where $v_p(=\omega/k_\|)$ and $v_g(=d\omega/dk_\|)$ respectively represent the phase and group velocities of the wave in the direction of **B**.

If parasitic resonances (see Section II.5) are neglected, as is usually permissible in the magnetosphere, the growth rate

$$\mathrm{Im}\,\omega \approx -(2\pi^3/\omega)(v_g v_p/c^2)|k_\|{}|^{-1}q_j^2(p_j/m_j v_r)^{2l}$$
$$\times\,B(s+1,\,l)[\omega + s(\omega - |\Omega_j|)]\,J_\perp(p_j^2/2m_j) \qquad (2.66)$$

thus follows from the interaction of electrons with the whistler mode ($j=e$) or from the interaction of protons with the proton-cyclotron mode ($j=i$). The growth rate is therefore positive at frequencies such that $0 < \omega < |s\Omega_j/(s+1)|$. Since (2.66) is based on the dynamics of an infinite homogeneous plasma, however, a positive value of Im ω is not synonymous with an instability that spontaneously generates appreciable wave intensity in the magnetosphere. Instability in this latter sense requires at least a small coefficient R of internal reflection, as in a maser, to prevent all the wave energy from escaping [52]. If the typical path length between the points of wave reflection is $\sim La$, then the condition for spontaneous wave generation (maser action) is

$$2La\,\mathrm{Im}\,\omega > |v_g \ln R|. \qquad (2.67)$$

This condition imposes an upper limit on the particle flux that the magnetosphere can stably contain. Instability in the sense of (2.67) generates wave energy that, by virtue of (2.47), causes the pitch angles of the excess particles to diffuse into the loss cone until (2.67) is no longer satisfied [52].

Limiting Flux. The upper limit on stably trapped particle flux is customarily expressed as a bound on the integral omnidirectional flux

$$I_{4\pi}(p_j^2/2m_j) = 4\pi \int_0^1 \sin^{2s}\alpha\,d(\cos\alpha)\int_{p_j^2/2m_j}^\infty J_\perp(p_j^2/2m_j)(p_j/p)^{2l}dE$$
$$= [\pi^{3/2}(p_j^2/m_j)\,\Gamma(s+1)/(l-1)\Gamma(s+\tfrac{3}{2})]\,J_\perp(p_j^2/2m_j) \qquad (2.68)$$

above the minimum particle energy with which a wave having $\text{Im}\,\omega > 0$ can resonate, $i.\,e.$, for a value of p_j given by $p_j^* = (m_j/k_{||})(\omega - |\Omega_j|)$, where $\omega = |s\Omega_j/(s+1)|$. The critical nonrelativistic particle energies for electrons resonant with (2.51a) and protons resonant with (2.51b) are therefore given respectively by

$$E^* = B^2/8\pi N_e(s+1)^2 s \qquad (2.69\,\text{a})$$

$$E^* = B^2/8\pi N_p(s+1)s^2. \qquad (2.69\,\text{b})$$

An estimate for the critical (maximum) value of $I_{4\pi}(E^*)$ is obtained from (2.66)—(2.68) by replacing $\omega + s(\omega - |\Omega_j|)$ with the value that maximizes $\text{Im}\,\omega/v_g$. This prescription tacitly ignores any frequency dependence of $\ln R$, $i.\,e.$, it is assumed that the internal-reflection coefficient is independent of wave frequency over the band of interest. It is thus estimated that $\text{Im}\,\omega/v_g$ peaks at

$$\omega + s(\omega - |\Omega_j|) \approx -(s/2)|\Omega_j|\left[(l-1)(v_p/v_g) + s\,l\right]^{-1} \qquad (2.70)$$

if $s\,l + (l-1)(v_p/v_g) \gg 1/2$.

If the particle spectrum is at least moderately steep $(l \gtrsim 4)$, then the value of $\text{Im}\,\omega/v_g$ does in fact descend sharply from a peak [where ω is given by (2.70)] to zero [where $\omega = |s\Omega_j/(s+1)|$], as required. The limiting flux is then given by

$$I_{4\pi}^*(E^*) \approx \frac{\{(v_p/v_g) + [s\,l/(l-1)]\}\,c\,B\Gamma(l+s+1)|\ln R|}{2\pi^{3/2}|q_j|(s+1)^2 s\Gamma(l)\,\Gamma(s+\tfrac{3}{2})L\,a}, \qquad (2.71)$$

with v_p/v_g evaluated at $\omega + s(\omega - |\Omega_j|) = 0$: thus $v_p/v_g = (s+1)/2$ for electrons and $1 + (s/2)$ for protons. For practical application, it is usual to insert apparently reasonable values of the various parameters $(e.\,g.,$ $v_p/v_g \sim 1$, $l \sim 4$, $s \sim 1/2$, $\ln R \sim -3)$ and to cite a common upper bound

$$I_{4\pi}^*(E^*) \sim (10^{11}/L^4)\,\text{cm}^{-2}\,\text{sec}^{-1} \qquad (2.72)$$

for the $equatorial$ integral omnidirectional flux of stably trapped particles (either electrons or protons, separately) exceeding the appropriate critical energy given by (2.69).

Observational evidence for such a limit on the flux of trapped electrons is shown in Fig. 22. For $s \sim 1/2$, the value of E^* given by (2.69a) approximates the magnetic energy per $plasma$ electron $(\approx 40\,\text{keV}$ if $N_e \approx 4\,\text{cm}^{-3}$ at $L = 5)$. Data points representing the omnidirectional flux of electrons greater than 40 keV in energy are distributed generally below $I_{4\pi}^*$ on the logarithmic scale; the flux exceeds $I_{4\pi}^*$ only in isolated (unstable) instances (see also Section IV.4).

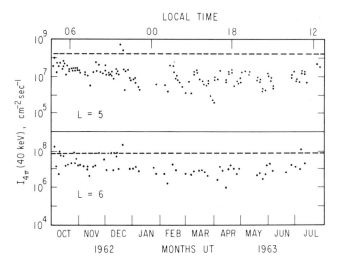

Fig. 22. Compilation of near-equatorial electron-flux measurements (magnetic latitudes $<30°$ at $L=5$ and $<15°$ at $L=6$) from Explorer 14 [52].

II.7 Weak Diffusion and Strong Diffusion

Although some degree of inelasticity is essential for wave growth or damping, cyclotron-resonant interactions of energetic (radiation-belt) particles are often preponderantly elastic in the sense that $D_{EE} \ll E|D_{xE}| \ll E^2 D_{xx}$. This property has been illustrated schematically in Fig. 19. The degree of inelasticity is of order $(\gamma_j \omega/\Omega_j)^2$ in the electromagnetic case, and the more energetic particles tend to resonate (Figs. 20—21) with the lower-frequency waves. Thus, the Jacobian given by (2.14) remains approximately applicable to the cyclotron-resonant pitch-angle diffusion of radiation-belt particles, and the equation

$$\frac{\partial \overline{f}}{\partial t} = \frac{1}{x\, T(y)} \frac{\partial}{\partial x} \left[x\, T(y) D_{xx} \frac{\partial \overline{f}}{\partial x} \right] \qquad (2.73)$$

approximates their dynamical behavior at constant energy.

Since $T(y)$ varies only moderately with $y \equiv (1-x^2)^{1/2}$, the solutions of (2.73) must somewhat reesemble those for symmetric diffusion in a cylinder. If D_{xx} is approximately constant in x, the decaying eigenfunctions of pitch-angle diffusion must then resemble the sequence

$\exp[-D_{xx}(\kappa_n/x_c)^2 t]J_0(\kappa_n x/x_c)$ in order to satisfy the boundary condition[18] that $\bar{f}=0$ at $x=x_c$ [43]. Here the κ_0 (≈ 2.40), κ_1 (≈ 5.52), κ_2 (≈ 8.65), etc., denote the roots of the Bessel function $J_0(x)$ in ascending sequence. This means that a suddenly injected distribution of trapped particles settles toward the lowest eigenmode of pitch-angle diffusion at the e-folding rate $\sim(\kappa_1^2-\kappa_0^2)(D_{xx}/x_c^2)$ upon removal of the particle source. Once the lowest eigenmode is attained, the pitch-angle distribution ceases to change its functional form, but proceeds to decay exponentially in time at a rate $\sim\kappa_0^2(D_{xx}/x_c^2)$. The characteristic lifetime of a particle against pitch-angle diffusion into the loss cone, *i.e.*, the e-folding time for the lowest eigenmode, thus exceeds by a factor ~ 4 the time required for attaining the lowest normal mode from an initially abnormal distribution of pitch angles (see Section IV.2).

A pitch-angle distribution of the form y^{2s} (see Section II.6) yields a decay rate of $4sD_{xx}$ at $x=0$ in either (2.35) or (2.73). If $x_c^2 s \sim 1$, this rate is quite comparable to the estimate $\kappa_0^2(D_{xx}/x_c^2)$ based on the cylinder analogy (see above). For comparison, the lowest eigenvalue of (2.20) is $(\pi^2/4)(D_{\xi\xi}/\xi_c^2)$ if $\bar{f}$ is required to vanish at $\xi=\pm\xi_c$. The numerical consistency of these estimates enables the particle lifetime to be characterized as $x_c^2/5D_{xx}$, within perhaps a factor of two.

The above considerations apply to a condition known as *weak diffusion*, in which the lifetime $x_c^2/5D_{xx}$ is sufficiently long compared to the bounce period $2\pi/\Omega_2$. The reason for this cautious interpretation is that the dense atmosphere is typically distant from the site of pitch-angle diffusion caused by a wave-particle interaction. Thus, the diffusing particles are unaware of a loss cone in momentum space (see Fig. 19) until they attempt to enter the dense atmosphere in the subsequent course of bounce motion. The requirement for applying (2.73) with the boundary condition that $\bar{f}=0$ at $x=\pm x_c$ amounts to the demand that

$$(\pi/\Omega_2)D_{xx} \ll (1-x_c)^2. \tag{2.74}$$

This means that a typical particle originating at $x=\pm x_c$ must be unable to wander (diffuse) across any substantial fraction of the loss-cone aperture during a single pass between mirror points. If (2.74) fails to hold, then the boundary condition that $\bar{f}=0$ at $x=\pm x_c$ applies only to upward-bound particles between the dense atmosphere and the site of pitch-angle diffusion, not to the entire bounce orbit. In such a case, the analysis based on (2.73) must take account of the substantial probability that a particle's pitch angle may wander into and back out of the loss

---

[18]This discussion is facilitated by the tacit assumption of a centered-dipole field, for which the loss-cone angle does not vary with longitude. The generalization to an offset-dipole model is indicated below.

cone (perhaps several times) during a single transit of the (typically equatorial) region where the wave-particle interaction occurs.

The foregoing considerations are based on the tacit assumption of a centered-dipole field. When the eccentricity of the dipole is taken into account, it is necessary to distinguish between the *bounce loss cone*, which contains particles that will *precipitate* (lose their energy to the dense atmosphere) within one bounce period, and the *drift loss cone*, whose aperture is given by (2.24). The aperture of the bounce loss cone varies with azimuth, and attains its maximum (identical with the aperture of the drift loss cone) near the South Atlantic "anomaly". Particles within the drift loss cone, but outside the local bounce loss cone, proceed to drift in azimuth but are doomed to precipitate somewhat prior to visiting the "anomaly".

An estimate for the bounce loss-cone angle $\cos^{-1} x_b$ follows from generalizing (2.24) to arbitrary longitude φ relative to the "anomaly", where $\varphi = \varphi_a$. The result, *viz.*,

$$1 - x_b^2 \approx [(a+h)/La]^3 [1 + 3 r_0 (La)^{-1/2} (a+h)^{-1/2} \cos(\varphi - \varphi_a)] \qquad (2.75)$$
$$\div [4 - (3/La)(a+h) - 3 r_0 (La)^{-3/2} (a+h)^{1/2} \cos(\varphi - \varphi_a)]^{1/2},$$

suggests a rather pronounced azimuthal modulation of the aperture of the bounce loss cone[19] at $L \lesssim 3$. As a result, particle precipitation tends to concentrate in azimuth near (slightly prior to) the "anomaly" in the absence of a counteracting variation of D_{xx} with φ (see Section IV.2). Since the meaning of (2.74) is somewhat ambiguous under azimuthal asymmetry, it is customary to regard *weak diffusion* as a condition applicable to any longitude at which

$$(\pi/\Omega_2)D_{xx} \ll (1 - x_b)^2. \qquad (2.76)$$

In the opposite extreme, if $D_{xx} \gg (\Omega_2/\pi)$, then the distribution $\bar{f}$ is immune from boundary conditions in x throughout the region in which pitch-angle diffusion occurs. On a single pass through this scattering region, a particle's probable pitch angle becomes thoroughly randomized over the unit sphere. This condition, known as *strong diffusion* [53], causes $\bar{f}$ to exhibit virtual isotropy in pitch angle within the scattering region. In the limit of strong diffusion, the decay rate of $\bar{f}$ is limited not by the magnitude of D_{xx} (of which $\partial \bar{f}/\partial t$ is actually independent), but rather by the solid angle of the bounce loss cone and by the bounce frequency $\Omega_2/2\pi$. Under strong diffusion, a fraction $1 - x_b$ of

---

[19]Thus, x_c is defined as the maximum value attained by the azimuthally varying parameter x_b.

the trapped particles will precipitate during each half bounce period, since the solid angle of either bounce loss cone is $2\pi(1-x_b)$. The loss rate is therefore given by

$$\lambda = -\partial \ln \bar{f}/\partial t = (\Omega_2/\pi)(1-x_b) \tag{2.77}$$

in the limit of strong diffusion. This limit represents an ultimate standard by which all diffusive precipitation mechanisms can be judged for effectiveness. Several storm-associated phenomena (*e.g.*, ring-current, relativistic-electron, and auroral precipitation) are actually observed to approximate the limit of strong diffusion [51, 52, 54].

III. Radial Diffusion

III.1 Violation of the Third Invariant

Whereas pitch-angle diffusion is customarily invoked as a loss mechanism for the radiation belts, diffusion in Φ is usually associated with creation of the belts. This is especially true of radial diffusion in which M and J are conserved, since particles then gain energy in the process of diffusing toward the earth from an external source (see below). Diffusion in Φ (radial diffusion) at constant M and J thus plays the dual role of injecting particles into the magnetospheric interior and accelerating the particles thereby injected to the energies observed.

In addition to particles that have entered from interplanetary space (and perhaps from the geomagnetic tail), the magnetosphere also contains protons and electrons born internally through the decay of albedo neutrons ejected from the upper atmosphere by energetic ($\gtrsim 100$ MeV) solar protons and galactic cosmic-ray particles colliding inelastically (in the nuclear sense) with gas atoms. These internal source mechanisms are known as SPAND and CRAND, respectively, for *solar-proton* (and *cosmic-ray*) *albedo neutron decay*. These sources (CRAND is about ten times as intense a particle source as SPAND) typically account for the presence of energetic protons and electrons in the inner zone, but radial diffusion plays an essential role in bringing about the observed spatial and spectral distribution of these particles [38]. In addition, radiation-belt particles may possibly experience *in situ* acceleration to high energies [44] through the absorption of plasma-wave energy. Such an event might easily be interpreted as an "injection" of the energetic particles into the magnetospheric interior (see Section IV.6).

Artificial radiation belts created by high-altitude nuclear detonations (1958—1963) once contributed substantially to the inner-zone particle population. These artificial belts, which had decayed to an intensity below that of the natural radiation by the year 1968, yielded some of the earliest measurements of a radial-diffusion coefficient for radiation-belt electrons in the magnetosphere.

In the outer zone, radial diffusion plays an all-important role in maintaining the level of trapped radiation. Direct observational evidence for the occurrence of third-invariant violation in the outer zone is shown in Fig. 23, which is a tracing of data obtained by instruments

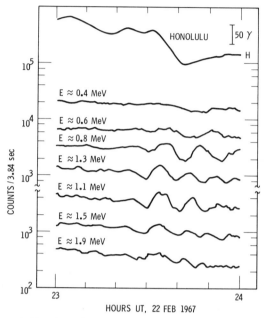

Fig. 23. Drift-periodic echoes in outer-zone electron fluxes, as observed on ATS 1 following a negative magnetic impulse [55] at 2330 UT (1330 LT).

on the geosynchronous equatorial satellite ATS 1 (longitude 150°W), together with the magnetogram (horizontal, or H, component) for the same time period (1300—1400 local time) from the ground-based station at Honolulu. The interpretation of Fig. 23 is that a negative magnetic impulse, presumably caused by a sudden decrease in solar-wind pressure at the magnetopause, propagates inward from the magnetopause and arrives at Honolulu several minutes after encountering the spacecraft [20]. Upon arrival at synchronous altitude, the impulse causes a simultaneous decrease of the electron flux observed in each of the seven energy channels. As time goes on, however, particles near the satellite at the arrival time of the impulse drift toward the night side, and electrons from the night side (where the negative impulse was less severe) drift to the azimuthal position of the satellite. This accounts for the recovery of the fluxes in each channel on a time scale of half the energy-dependent drift period. The relative minimum in flux recurs with the return (to the day side) of those particles most severely influenced by the impulse.

[20]This delay time is in accord with the time required for a rarefactional (magnetosonic) impulse to travel the required distance of 5.6 earth radii at approximately the Alfvén speed.

These *drift-periodic echoes* in the outer-zone electron flux persist well after the passage of the magnetic impulse initiating them. Moreover, the fact that each energy channel "oscillates" at its own characteristic drift frequency is convincing evidence for drift-phase organization of the particles, which therefore (*cf.* Section II.1) have been dispersed with respect to $|\Phi|$ ($\equiv 2\pi a^2 B_0/L$)[21]. The nonvanishing energy bandwidth of each detection channel corresponds to a drift-frequency bandwidth that thoroughly phase-mixes the observations on a time scale of three or four drift periods. Particles initially differing in both φ_3 and energy retain their separate identities, but the detectors can no longer distinguish among them.

The practical fact of phase mixing, and the fact that consecutive sudden impulses are statistically uncorrelated on the drift time scale, provide the essential degree of randomness that makes it appropriate to speak of third-invariant violation in terms of diffusion with respect to Φ. At constant M and J, *i.e.*, with pitch-angle diffusion neglected, the radial-diffusion equation

$$\frac{\partial \bar{f}}{\partial t} = \frac{\partial}{\partial \Phi}\left[D_{\Phi\Phi}\frac{\partial \bar{f}}{\partial \Phi}\right] = L^2 \frac{\partial}{\partial L}\left[\frac{1}{L^2} D_{LL}\frac{\partial \bar{f}}{\partial L}\right] \qquad (3.01)$$

follows directly from (2.01), since $D_{LL} \equiv (dL/d\Phi)^2 D_{\Phi\Phi}$. The distribution function $\bar{f}$ is equal to $\bar{J}_{\perp}/p^2$, evaluated on a surface generated by the mirror points of ions or electrons having in common their values of M and J.

In a dipole field this surface coincides with the equatorial plane ($\theta = \pi/2$) for particles having $J = 0$. For $J \neq 0$ the mirror-point surface satisfies the equation

$$[y/Y(y)]^2 = 8 m_0 B_0 a^2 (M/J^2 L) \equiv B_0 a^2/K^2 L, \qquad (3.02)$$

where y is related by (1.25) to the mirror colatitude θ_m. With the aid of (3.02) and (1.31), the variation of y with L at constant M and J is plotted in Fig. 24 for selected values of y_7 (the value of y at $L=7$). The $L=7$ shell is often used as a reference location in radiation-belt theory because it is quite near the outer boundary of stable trapping, and therefore adjacent to a possibly important source of moderately energetic particles (*i.e.*, solar cosmic rays that have entered the magnetosphere). A secondary reason for the popularity of $L=7$ as a reference shell [40] is that an equatorially mirroring particle's nonrelativistic

---

[21]This follows from Liouville's theorem, since $J_3/2\pi$ ($\equiv q\Phi/2\pi c$) is canonically conjugate to the drift phase φ_3.

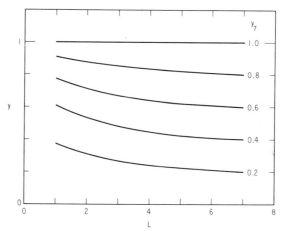

Fig. 24. Systematic variation of y (sine of equatorial pitch angle) with L at constant M and J, applicable to radial diffusion caused by magnetospheric impulses.

kinetic energy $p^2/2m_0$ (measured in keV) at $L=7$ roughly approximates the particle's first invariant M (measured in MeV/gauss).

Except at the end-points $y=0$ and $y=1$, there is a systematic inverse variation (not proportionality) between y and L during radial diffusion at constant M and J. This variation does not constitute pitch-angle diffusion, but rather is an interesting property incidental to radial diffusion. The change in particle energy during diffusion in L can be deduced from the identity $p^2 = 2m_0 M B_0/L^3 y^2$ if $0<y\le1$, or from $p=J/2LaY(y)$ if $0\le y<1$. It follows that p^2 varies more strongly than L^{-2}, but more weakly than L^{-3}, in the interval $0<y<1$.

The type of radial diffusion that conserves both M and J [58] can be caused by magnetic sudden impulses (as illustrated in Fig. 23), by substorm-associated impulses of the convection electrostatic field, and by other magnetospheric disturbances operating on a similar time scale (~ 100 sec). In each case the affected particles yield a bounce-averaged response, since the rise time of the impulse (~ 100 sec) is typically much longer than $2\pi/\Omega_2$ (~ 1 sec). On the other hand, the drift periods of many radiation-belt particles (~ 500 sec in Fig. 23) are not extremely long compared to the rise time of a typical sudden impulse, and so a frequency-spectral treatment of impulses is definitely in order. In such a treatment, a particle responds resonantly to Fourier components located at harmonics (including the fundamental) of its drift frequency, although the impulses themselves are hardly oscillatory in character.

In addition to the type of radial diffusion that conserves both M and J, it is possible to conceive of mechanisms that fail to preserve the first two invariants while violating Φ. Such mechanisms may involve particle collisions or bounce- and cyclotron-resonant interactions with magnetospheric waves. Radial diffusion mechanisms that violate M and/or J often lack the ability to energize particles efficiently in the process, and they generally play a less certain role than sudden impulses in the overall picture of radiation-belt dynamics.

III.2 Magnetic Impulses

In the magnetic-field model specified by (1.45), sudden impulses in $\mathbf{B}$ correspond to sudden changes in b, the geocentric stand-off distance to the subsolar point on the magnetopause. The stand-off distance b is governed, according to (1.43), by the momentum flux of the solar wind. An encounter with the plasma ejected by a solar flare, for example, can lead to a sudden contraction and/or expansion of the magnetosphere. A decrease in b that is sudden on the drift time scale represents a sudden contraction of the magnetosphere. This contraction consists of both an azimuthally symmetric compression of $\mathbf{B}$ (the B_1 term) and an azimuthally asymmetric distortion of $\mathbf{B}$ (the B_2 term). The symmetrical compression, which is easily identified from the magneto-grams of ground based ($r=a$) observatories, is adiabatic to the trapped particles. All drift phases φ_3 respond identically to the symmetric part of the sudden impulse, and so this part is reversible. It conserves Φ and produces no radial diffusion.

Induced Electric Field. The accompanying asymmetric distortion (the B_2 term) is not easily distinguished at $r=a$, where it is small in magnitude. However, this part of the impulse does violate the third invariant, thereby producing drift echoes (Fig. 23) and radial diffusion. A sudden impulse in $\mathbf{B}$ affects the geomagnetically trapped particles by virtue of an induced electric field $\mathbf{E}$, which may be calculated term by term from a field expansion [29] of the form [cf. (1.46)]

$$E_r(r, \theta, \varphi; t) = \sum_{lmn} E_r(l, m, n; t)(r/b)^n \sin^l \theta \sin m\varphi \qquad (3.03\,a)$$

$$E_\theta(r, \theta, \varphi; t) = \sum_{lmn} E_\theta(l, m, n; t)(r/b)^n \cos \theta \sin^l \theta \sin m\varphi \qquad (3.03\,b)$$

$$E_\varphi(r, \theta, \varphi; t) = \sum_{lmn} E_\varphi(l, m, n; t)(r/b)^n \sin^l \theta \cos m\varphi . \qquad (3.03\,c)$$

If the Maxwell relation $c\nabla \times \mathbf{E} = -(\partial \mathbf{B}/\partial t)$, written out in its three components, is applied to (1.46) and (3.03), the time-dependent (but position-independent) coefficients of $[(r/b)^n \sin^{l+1}\theta \cos m\varphi]$ and $[(r/b)^n \cos \theta$

$\times \sin^l \theta \sin m\varphi$] can be isolated to yield the relationships

$$(n+1) E_\varphi(l, m, n; t) = m E_r(l+1, m, n; t)$$
$$+ (a/c)(b/a)^n \dot{B}_\theta(l, m, n-1; t) \qquad (3.04a)$$

$$(n+1) E_\theta(l, m, n; t) = (l+1) E_r(l+1, m, n; t)$$
$$- (a/c)(b/a)^n \dot{B}_\varphi(l, m, n-1; t), \qquad (3.04b)$$

where $\dot{\mathbf{B}} = \partial \mathbf{B}/\partial t$. The third component of $c\nabla \times \mathbf{E} = -(\partial \mathbf{B}/\partial t)$ is redundant, since $\nabla \cdot \mathbf{B} = 0$.

One more condition on $\mathbf{E}$ must be specified in order to solve (3.04). It is customary to state this subsidiary condition as $\mathbf{E} \cdot \mathbf{B} = 0$ [32]. Such a statement is usually justified by an appeal to the cold plasma which is assumed to fill the magnetosphere. The cold plasma serves to short-circuit each field line, in which case the impulsively expanding $(db/dt > 0)$ or contracting $(db/dt < 0)$ magnetospheric medium is governed by the laws of magnetohydrodynamics (MHD). Since the impulse therefore propagates through the magnetosphere at approximately the Alfvén speed, the field model summarized by (1.46) admittedly violates the principle of causality on time scales shorter than $\sim b/c_A$. For drift periods exceeding a few minutes, however, the arrival time of the impulse at any L shell is *practically* independent of φ_3, and this condition permits the simplified (instantaneous-response) model to be used for the time-varying $\mathbf{B}$ field.

For the magnetic-field model given by (1.48), application of $\mathbf{E} \cdot \mathbf{B} = 0$ to (3.04) yields the recursion relation [29]

$$[(2n+l+2)/(n+1)] E_r(l, m, n; t)$$
$$= (B_1/B_0)[(n-l-2)/(n-2)] E_r(l, m, n-3; t)$$
$$+ (B_2/B_0)[(l-n+2)/(n-3)] E_r(l-1, m-1, n-4; t)$$
$$+ (B_2/B_0)[(l-n-2)/(n-3)] E_r(l-1, m+1, n-4; t)$$
$$- (B_2/2B_0)[(l-m+2)/(n-3)] E_r(l+1, m-1, n-4; t)$$
$$- B_2/2B_0)[(l+m+2)/(n-3)] E_r(l+1, m+1, n-4; t) \qquad (3.05)$$
$$- (4/3c) B_2(a/b)^3 (db/dt)\delta_{l1} \delta_{m1} \delta_{n2}$$
$$+ (1/6c) B_2(B_1/B_0)(a/b)^3 (db/dt)\delta_{l1} \delta_{m1} \delta_{n5},$$

where the Kronecker symbol δ_{ij} is equal to unity (rather than zero) only if $i = j$. Closure of the recursion formula is achieved by requiring that the $\mathbf{E}$ induced by db/dt remain finite in the limit $r = 0$. This requirement forces $E_r(l, m, n; t)$ to vanish if $n < 0$. From this starting point it is possible to generate all the coefficients $E_r(l, m, n; t)$ by means of (3.05). The nonvanishing $E_r(l, m, n; t)$ of lowest order n is $E_r(1, 1, 2; t) = -(4/7c)(a/b)^3 (db/dt) B_2$.

The recursion relation yields definite algebraic values for several coefficients which have $m=0$, and which therefore ostensibly can have no physical significance [see (3.03a)]. Such coefficients that multiply zero (in the form of $\sin m\varphi$) are ignored. A correct implementation of (3.05) thus leads to a unique set of electric-field coefficients $E_r(l,m,n;t)$, where $m>0$, from which $E_\theta(l,m,n;t)$ and $E_\varphi(l,m,n;t)$ are obtainable by means of (3.04). All nonvanishing electric-field coefficients for which $n \leq 10$ are listed in Table 7.

Table 7. Electric-Field Coefficients

$$E_r(1,1,2;t) = -(4/7)B_2(a^3\dot{b}/b^3c)$$
$$E_r(1,1,5;t) = -(9/91)(B_1/B_0)B_2(a^3\dot{b}/b^3c)$$
$$E_r(2,2,6;t) = (1/6)(B_2/B_0)B_2(a^3\dot{b}/b^3c)$$
$$E_r(1,1,8;t) = -(135/3458)(B_1/B_0)^2 B_2(a^3\dot{b}/b^3c)$$
$$E_r(2,2,9;t) = (25/273)(B_2/B_0)^2 B_1(a^3\dot{b}/b^3c)$$
$$E_r(1,1,10;t) = -(11/483)(B_2/B_0)^2 B_2(a^3\dot{b}/b^3c)$$
$$E_r(3,1,10;t) = -(11/210)(B_2/B_0)^2 B_2(a^3\dot{b}/b^3c)$$
$$E_r(3,3,10;t) = -(11/210)(B_2/B_0)^2 B_2(a^3\dot{b}/b^3c)$$

$$E_\theta(0,1,2;t) = (8/7)B_2(a^3\dot{b}/b^3c)$$
$$E_\theta(0,1,5;t) = -(3/182)(B_1/B_0)B_2(a^3\dot{b}/b^3c)$$
$$E_\theta(1,2,6;t) = (1/21)(B_2/B_0)B_2(a^3\dot{b}/b^3c)$$
$$E_\theta(0,1,8;t) = -(15/3458)(B_1/B_0)^2 B_2(a^3\dot{b}/b^3c)$$
$$E_\theta(1,2,9;t) = (5/273)(B_2/B_0)^2 B_1(a^3\dot{b}/b^3c)$$
$$E_\theta(0,1,10;t) = -(1/483)(B_2/B_0)^2 B_2(a^3\dot{b}/b^3c)$$
$$E_\theta(2,1,10;t) = -(1/70)(B_2/B_0)^2 B_2(a^3\dot{b}/b^3c)$$
$$E_\theta(2,3,10;t) = -(1/70)(B_2/B_0)^2 B_2(a^3\dot{b}/b^3c)$$

$$E_\varphi(1,0,1;t) = (3/2)B_1(a^3\dot{b}/b^3c)$$
$$E_\varphi(0,1,2;t) = (8/7)B_2(a^3\dot{b}/b^3c)$$
$$E_\varphi(2,1,2;t) = -(8/3)B_2(a^3\dot{b}/b^3c)$$
$$E_\varphi(0,1,5;t) = -(3/182)(B_1/B_0)B_2(a^3\dot{b}/b^3c)$$
$$E_\varphi(1,2,6;t) = (1/21)(B_2/B_0)B_2(a^3\dot{b}/b^3c)$$
$$E_\varphi(0,1,8;t) = -(15/3458)(B_1/B_0)^2 B_2(a^3\dot{b}/b^3c)$$
$$E_\varphi(1,2,9;t) = (5/273)(B_2/B_0)^2 B_1(a^3\dot{b}/b^3c)$$
$$E_\varphi(0,1,10;t) = -(1/483)(B_2/B_0)^2 B_2(a^3\dot{b}/b^3c)$$
$$E_\varphi(2,1,10;t) = -(1/70)(B_2/B_0)^2 B_2(a^3\dot{b}/b^3c)$$
$$E_\varphi(2,3,10;t) = -(1/70)(B_2/B_0)^2 B_2(a^3\dot{b}/b^3c)$$

Response of Trapped Particles. This analytical representation of the **E** field induced by a time-varying model **B** field is especially useful for following the response of trapped particles to a magnetic impulse. Each particle experiences an electric drift at velocity

$$\mathbf{v}_d = (c/B^2)\mathbf{E} \times \mathbf{B} \qquad (3.06)$$

in addition to its gradient, curvature, and other (cf. Section III.6) drifts. As a consequence, the particle may change its value of $L (\equiv 2\pi a^2 B_0|\Phi|^{-1})$.

Except for particles mirroring at the equator, the bounce average required in applying (3.06) is quite onerous. A rather different approach, based on (1.77b), is more expedient for calculating the radial-diffusion coefficient D_{LL} to lowest order in (B_2/B_0), for arbitrary mirror latitude.

The more expedient approach is based on the fact that v_d, as given by (3.06), can be identified as the local velocity of a field line if the $\mathbf{E}$ field induced by $\partial \mathbf{B}/\partial t$ is everywhere perpendicular to $\mathbf{B}$. In other words, if only the $\mathbf{E} \times \mathbf{B}$ drift is considered, the particle remains on its original field line, as identified by the label L_d. The proof that field-line motion can be traced in this manner follows from the identity

$$B^2(d L_d/d t) = B^2(\partial L_d/\partial t) + B^2 v_d \cdot \nabla L_d$$
$$= B^2(\partial L_d/\partial b)(d b/d t) + c \mathbf{E} \times \mathbf{B} \cdot \nabla L_d = 0, \qquad (3.07)$$

which can be verified with the aid of Table 7 to each order in ε_1 and ε_2 (see Section I.7). The degree of accuracy inherent in (1.69b), which implies

$$L_d \approx (r/a \sin^2 \theta)[1 + (B_1/2 B_0)(r/b)^3$$
$$- (2 B_2/21 B_0 \sin \theta)(r/b)^4(7 \sin^2 \theta - 3) \cos \varphi], \qquad (3.08)$$

is adequate to verify (3.07) in first and second order[22]. A more extensive proof (to higher order) is not required here, but could easily be generated [32].

Whereas the $\mathbf{E} \times \mathbf{B}$ drift induced by a time variation of b yields no immediate change in a particle's L_d coordinate, the gradient-curvature drift does. According to (1.77b) this change is of the form

$$d L_d/d t \approx - \dot{\varphi}(B_2/252 B_0) L_d^5(a/b)^4 [Q(y)/D(y)] \sin \varphi. \qquad (3.09)$$

The coordinate L_d to which the particle would return at $\varphi = \pm \pi/2$ in a magnetosphere frozen in time $(db/dt=0)$ properly labels the drift shell in the sense that

---

[22]The field-line label L_d defined by (1.51b) is conceptually useful only if higher internal geomagnetic multipoles, which would dominate the dipole as r approaches zero, are neglected. More generally, a field line may be labeled by the point at which it is anchored in the surface of an ideally conducting solid (of which the earth is an adequate example on the time scales of interest). If due account is taken of currents thereby induced on the surface of this conducting solid, it can be shown that the $\mathbf{E} \times \mathbf{B}$ drift induced by $\partial \mathbf{B}/\partial t$ impels a particle to remain attached to its original field line, as identified by the coordinates of its foot on the conducting surface [32]. In a model such as (1.45) there is no provision for currents at $r=a$, and the conducting solid degenerates to a point at the origin $(r=0)$.

$$|\Phi| = \int_0^{2\pi} \int_{r_e(\varphi)}^{\infty} B_0 (a/r)^3 r \, dr \, d\varphi$$

$$- \int_0^{2\pi} \int_0^{r_e(\varphi)} [B_1(a/b)^3 - B_2(a/b)^3(r/b)\cos\varphi] r \, dr \, d\varphi$$

$$= [2\pi a^2 B_0/L_d(\pi/2)]\{1 + O(\varepsilon_2^2) + \cdots\}. \tag{3.10}$$

For convenience, the third invariant has been evaluated using the equatorial plane ($\theta = \pi/2$) of the magnetosphere, in which case $\mathbf{B}$ points in the $-\hat{\theta}(+\hat{z})$ direction and has a magnitude given by (1.45b). Since the end result of (3.10) has no correction terms of order ε_1 or ε_2, the definition $L \equiv L_d(\pm\pi/2)$ suffices for a calculation of D_{LL} to lowest order.

Diffusion Coefficient. Since $L \equiv L_d(\pm\pi/2)$, it follows from (1.77b) that

$$L = L_d\{1 - (B_2/252 B_0)(L_d a/b)^4 [Q(y)/D(y)]\cos\varphi\}. \tag{3.11}$$

The instantaneous shell parameter L thus changes at a rate

$$dL/dt = (B_2/63 a B_0)(L a/b)^5 (db/dt)[Q(y)/D(y)]\cos\varphi \tag{3.12}$$

to lowest order in ε_1 and ε_2. The radial diffusion coefficient $D_{LL} \equiv (1/2\tau)\langle(\Delta L)^2\rangle$ is obtained by integrating (3.12) over an interaction time $\tau \gg 2\pi/\Omega_3$, during which $\cos\varphi = \cos(\Omega_3 t + \varphi_3)$. It is convenient to express the result in terms of $\mathcal{B}_z(\omega/2\pi)$, which is defined as the spectral density function of $B_1(a/b)^3$. The procedure for obtaining D_{LL} is much the same as that used in Section II.4, and the result [56, 57] is

$$D_{LL} = 2\Omega_3^2(B_2/756 B_1 B_0)^2 L^{10}(a/b)^2 [Q(y)/D(y)]^2 \mathcal{B}_z(\Omega_3/2\pi). \tag{3.13}$$

When radial diffusion is caused by magnetic impulses, the energy dependence of D_{LL} is contained entirely in Ω_3. If the impulses rise sharply and decay slowly (like a step function) on the drift time scale, then $\mathcal{B}_z(\Omega_3/2\pi)$ is proportional to Ω_3^{-2} and *all energy dependence disappears*. At sufficiently high energies, the particle drift period becomes somewhat comparable to the rise time of an impulse (see Fig. 22), and this range of drift frequencies finds $\mathcal{B}_z(\Omega_3/2\pi)$ falling more sharply than Ω_3^{-2}. It follows that D_{LL} ultimately decreases somewhat with increasing energy. At constant M and J, however, the drift frequency decreases with increasing L. Thus, any inverse dependence of D_{LL} on particle energy tends to strengthen the L dependence of the radial-diffusion coefficient.

The dependence of D_{LL} on equatorial pitch angle is contained primarily in the factor $[Q(y)/D(y)]^2$. This factor, as approximated within $\sim 1\%$ by means of (1.36) and (1.79), varies by nearly an order of magnitude between $y=1$ and $y=0$ [56, 57]. The function $Q(y)/180\,D(y)$ is plotted in Fig. 25. At energies sufficiently high that the drift period is comparable to the rise time of a magnetic impulse, this extreme variation of D_{LL} with x is slightly moderated by the fact that particles having $x \sim 1$ drift more slowly in azimuth than those for which $x=0$ [see (1.35)]. At a given energy, of course, this variation of Ω_3 with x is very weak.

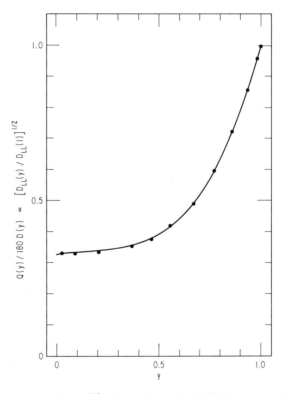

Fig. 25. Variation of $(D_{LL})^{1/2}$ with y for radial diffusion caused by magnetic sudden impulses. Data points have been determined by numerical computation [19]. Solid curve is analytical approximation [66] based on (1.36) and (1.79).

In summary, the radial-diffusion coefficient caused by magnetic impulses that rise sharply and decay slowly on the drift time scale is virtually independent of energy. For particles mirroring at the equator, the coefficient is given [29] by

$$D_{LL} = 2\Omega_3^2 (5B_2/21\,B_1\,B_0)^2 L^{10}(a/b)^2 \,\mathscr{B}_z(\Omega_3/2\pi) \qquad (3.14)$$

since $Q(1)=180D(1)$. In the case that $\mathscr{B}_z(\Omega_3/2\pi)$ falls off as Ω_3^{-2}, there is no energy dependence in D_{LL}. Thus, the diffusion coefficient depends on y through the factor $[Q(y)/D(y)]^2$ in (3.13), and on L through the factor L^{10}. But since y and L are related via (3.02), the factor $[Q(y)/D(y)]^2$ exhibits an inverse variation with L. Except at $y\equiv0$ and $y\equiv1$, this factor tends to moderate the variation of D_{LL} with L. With the aid of Fig. 25, it is possible to evaluate the ratio of D_{LL} at any L to D_{LL} at $L=7$ and $y=1$ for selected values of y_7, under the assumption that $\omega^2 \mathscr{B}_z(\omega/2\pi)$ is a constant. The results, plotted in Fig. 26, are principally of interest for high-energy protons and helium ions, which may in fact escape pitch-angle diffusion during their period of residence in the magnetosphere.

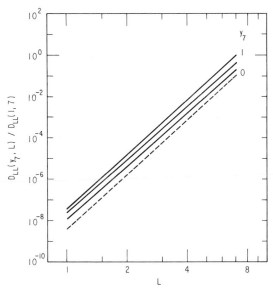

Fig. 26. Radial variation of D_{LL} driven by magnetic impulses, for selected values of y at $L=7$ ($y_7=0$, 0.4, 0.6, 1.0). Dashed line ($y_7=0$) is not realized in practice, because of the loss cone.

The magnitude of the spectral density function $\mathscr{B}_z(\omega/2\pi)$ is likely to vary with magnetic activity (as measured by an index such as K_p or D_{st}), and so the otherwise arbitrary interaction time τ used in this chapter is limited by the time scale of several days characteristic of changes in magnetospheric "weather". However, if the observations of $\bar{f}$ are sufficiently coarse-grained as to average over genuine temporal variations of D_{LL}, it may be possible to identify a mean diffusion coefficient applicable to a much longer time interval (see Chapter V). Particle lifetimes in the inner proton belt are even long enough (see Fig. 14, Section II.2) to permit averaging D_{LL} and $\bar{f}$ over a large number of solar cycles.

III.3 Electrostatic Impulses

Abrupt temporal changes in the electrostatic potential associated with plasma convection are characteristic of geomagnetic storms and magnetospheric substorms. Impulses of this type can be represented by a time-dependent coefficient E_c in (1.52). For particles of radiation-belt energy $(W \gg |q V_e|)$, the drift-shell asymmetry caused by the *mean* value of E_c can be neglected in the calculation of D_{LL} to lowest order in $q E_c a L_d / W$. Moreover, the magnetic field is assumed to be given by (1.16). In this situation it is evident from (1.66) that $|\Phi| = 2\pi a^2 B_0 / L_d(0)$, with no first-order correction in $q E_c a L_d / W$. Thus, the instantaneous shell parameter L is given in lowest order by

$$
\begin{aligned}
L &= L_d \{ 1 - [\gamma/(\gamma^2 - 1)](q E_c a L_d / 3 m_0 c^2)[T(y)/D(y)] \sin \varphi \} \\
&= L_d \{ 1 - (E_c/B_0)(c/\Omega_3 a) L_d^2 \sin \varphi \} ,
\end{aligned}
\tag{3.15}
$$

in view of (1.35). Since (1.65a), to the order of accuracy inherent in (3.15), implies that

$$
d L_d / dt = \dot{\varphi}(E_c/B_0)(c/\Omega_3 a) L_d^2 \cos \varphi
\tag{3.16}
$$

for a particle drifting in azimuth under the influence of (1.52), it follows that

$$
d L / dt = -(\dot{E}_c/B_0)(c/\Omega_3 a) L_d^3 \sin \varphi
\tag{3.17}
$$

for this particle. With E_c represented as the sum of purely temporal Fourier components [cf. (2.28)], the particle selects that component for which $\omega = \Omega_3$ after an interaction time $\tau \gg 2\pi/\Omega_3$.

The diffusion coefficient $D_{LL} \equiv (1/2\tau)\langle(\Delta L)^2\rangle$ obtained from (3.17) by the methods of Section II.4 is given by

$$
D_{LL} = 2(c/4 a B_0)^2 L^6 \mathscr{E}_c(\Omega_3/2\pi),
\tag{3.18}
$$

where $\mathscr{E}_c(\omega/2\pi)$ is the spectral density function of E_c (see Section I.6). A particle's energy and equatorial pitch angle enter (3.18) only via Ω_3. For electrostatic impulses that rise sharply and decay slowly on the drift time scale, the spectral density $\mathscr{E}_c(\Omega_3/2\pi)$ falls as Ω_3^{-2}. In this case, the functional form of D_{LL} is

$$
\begin{aligned}
D_{LL} = 2(q a/24 B_0)^2 L^{10} [T(y)/D(y)]^2 (y^2/M)^2 \\
\times [1 + (2 M B_0/m_0 c^2 y^2 L^3)] \omega^2 \mathscr{E}_c(\omega/2\pi) ,
\end{aligned}
\tag{3.19}
$$

where $\omega/2\pi$ is any frequency whose reciprocal lies well between the rise and decay times of the typical electrostatic impulse.

It is conventional to compare spatially coincident magnetospheric fluxes of different ionic species at common kinetic energy per nucleon, *i.e.*, at common particle velocity. This convention greatly simplifies the comparative analysis of collisional effects (see Section II.2). The ions of interest are typically nonrelativistic, and so the comparison applies essentially at common y, L, and M/A, where A is the number of nucleons in the ion. The respective electrostatic radial-diffusion coefficients thus scale as $(q/A)^2$. The magnitudes of D_{LL} for $H^+ : He^{++} : He^+$ therefore scale as $16:4:1$. When coupled with the expectation that magnetospheric helium nuclei (originally interplanetary alpha particles) spend up to half their radiation-belt lifetimes as He^+ by virtue of charge exchange (see Section II.2), this property of electrostatic diffusion provides a possibly interesting explanation [40] for the observational fact (see Section IV.5) that ratios of helium-ion flux to proton flux (often abbreviated α/p and He^+/p) at common E/A in the magnetosphere (well off the equator) are orders of magnitude smaller than the α/p ratio in the solar wind [59].

Among particles of the same species, the diffusion coefficient given by (3.19) is rather sensitive to particle energy, but notably insensitive

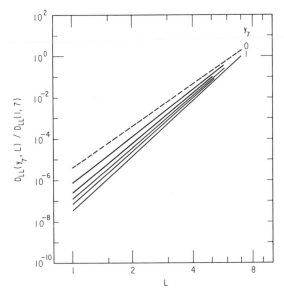

Fig. 27. Radial variation of D_{LL} driven by electrostatic impulses, for nonrelativistic particles having a common value of E at $L=7$ and selected values of y at $L=7$ ($y_7=0$, 0.2, 0.4, 0.6, 0.8, 1.0). Dashed line ($y_7=0$) is not realized in practice.

to equatorial pitch angle. Neither energy nor pitch angle is invariant during radial diffusion at constant M and J, however. Thus, a proper comparison should follow the spirit of Fig. 26, wherein particles are distinguished according to their values of y at $L=7$ in a dipole field. In electrostatic diffusion it is logical to compare particles having a common kinetic energy E_7 at $L=7$, i.e., a common value of M/y_7^2. The derivation of (3.19) assumes that $\omega^2 \mathscr{E}_c(\omega/2\pi)$ is constant for all frequencies of interest. If attention is limited to nonrelativistic particles, such as radiation-belt ions, the variation of D_{LL} with L is that of $L^{10}[T(y)/D(y)]^2(y/y_7)^4$. With the aid of Fig. 24, which indicates the variation of y with L for selected values of y_7, the ratio of D_{LL} at arbitrary L to D_{LL} at $L=7$ and $y=1$ has been evaluated for these selected values of y_7. The result is shown in Fig. 27. The common value of E_7 that forms the basis of this comparison is assumed to be such that $270 E_7 \ll m_0 c^2$ in order to justify using the nonrelativistic form of (3.19) down to $L \approx 1.08$, where the dense atmosphere terminates the inner belt (see Section II.2).

Spectral Density. Extrapolation of (3.19) to ring-current energies and below is forbidden on a variety of grounds. Consider (for example) a random sequence of impulses, each consisting of an ideally instantaneous jump from E_c to $E_c + \Delta E_c$, followed by an exponential decay to E_c with an e-folding time τ_d. The spectral density $\mathscr{E}_c(\omega/2\pi)$ is then given[23] by

$$\mathscr{E}_c(\omega/2\pi) = \frac{2(\tau_d^2/\tau)\Sigma(\Delta E_c)^2}{1+\omega^2\tau_d^2}, \qquad (3.20)$$

where $\omega > 0$ and $\Sigma(\Delta E_c)^2$ denotes the sum of the squares of all sudden increments in E_c initiated within an arbitrarily long (but statistically homogeneous) time interval of duration τ. The validity of (3.19) thus requires $\omega^2\tau_d^2 \gg 1$ at $\omega = |\Omega_3|$. It is presumed that $\tau_d \sim 2\,\mathrm{hr}$, and that most radiation-belt particles therefore comply with the conditions of (3.19). At ring-current (hot-plasma) energies and below, it is essential to reconsider the radial-diffusion problem in terms of (1.65), without making the simplifying approximation that $W \gg |q V_e|$.

On the other hand, the spectral density $\mathscr{E}_c(\Omega_3/2\pi)$ falls more rapidly than Ω_3^{-2} for particles having drift periods comparable to or smaller than the rise time of a typical impulse. The approximation of a vanishing rise time, as used in (3.20), is appropriate only if the particles of interest

---

[23]The spectral density function given by (3.20) has been constructed in a manner easily generalized to other types of impulses. For example, if E_c is replaced by $B_1(a/b)^3$ in (3.20), the result is a formula for $\mathscr{B}_z(\omega/2\pi)$, the spectral density function for magnetic sudden impulses (Section III.2).

have drift periods well in excess of the true rise time. Moreover, the magnitude of $\mathscr{E}_c(\omega/2\pi)$ is likely to vary with the level of geomagnetic activity, as measured by an index such as K_p or D_{st} (cf. Section III.2).

Harmonic Resonances. Large-scale electrostatic fields in the magnetosphere presumably may fluctuate in other than the simple mode assumed in (3.17). For example, it may be impossible to represent the *fluctuating* $V_e(r,\theta,\varphi;t)$ as in (1.52), but quite reasonable to represent it as

$$V_e(r, \theta, \varphi; t) = \sum_m E_m(L_d, t)\, a\, L_d \sin(m\,\varphi + \psi_m)\,. \tag{3.21}$$

The case culminating in (3.18) is included in (3.21) if $E_1(L_d,t)=E_c(t)$ and $\psi_1 = 0$. The more general expression for $V_e(r,\theta,\varphi;t)$, however, yields a diffusion coefficient of the form [56]

$$D_{LL} = 2(c/4\,a\,B_0)^2\, L^6 \sum_m m^2\, \mathscr{E}_m(L, m\,\Omega_3/2\,\pi)\,, \tag{3.22}$$

where $\mathscr{E}_m(L,\omega/2\pi)$ is the spectral density function of $E_m(L_d,t)$. The additional spatial structure present in (3.21) thus partially transfers the burden of causing radial diffusion to the higher harmonics of the drift frequency. The entry of these higher harmonics is reminiscent of a similar effect in bounce-resonant pitch-angle diffusion (see Section II.4) and occurs for an analogous reason (lack of positive long-range spatial correlation). Magnetospheric observations of a fluctuating electrostatic field must therefore be treated with caution in terms of extracting a diffusion coefficient, unless the extent of spatial coherence is known.

III.4 Bounce Resonance

Resonance of an MHD or electrostatic wave with harmonics of a particle's bounce frequency has been invoked previously (see Section II.4) as a mechanism for pitch-angle diffusion. There it was noted that confinement of the electric-field perturbation **e** to a meridional plane would prevent contamination by radial-diffusion effects. Conversely, a component of **e** in the azimuthal direction provides for the possibility of radial diffusion. The case of an electrostatic wave (for which **e** is parallel to **k**) propagating *purely* in the azimuthal direction (see also Section III.5) is essentially covered by (3.21) and (3.22). The resonance condition is found to be $\omega = m\Omega_3$. Even if m is a very large number ($\sim\varepsilon^{-1}$ for a resonant particle), the resonance condition is unaffected by the bounce motion if **k** is everywhere normal to **B** for an electrostatic wave.

If, however, the electrostatic wave is such that field lines are *not* equipotentials, *e. g.*, as in (2.37), then the condition for resonance takes on the form [64]

$$\omega - m\Omega_3 - l\Omega_2 = 0. \tag{3.23}$$

Just as in (2.33), the various values of l enter the diffusion coefficient weighted by $J_l^2(k_\parallel px/\gamma m_0 \Omega_2)$, which is small if the order (l) is much larger than the argument $(k_\parallel px/\gamma m_0 \Omega_2)$. If $m \ll |\varepsilon|^{-1}$, it may be instructive to define an azimuthal wavenumber $k_\varphi \equiv (m/r)\csc\theta$ and a bounce-averaged particle drift velocity $v_\varphi = \Omega_3 r \sin\theta$ in the dipole field. The resonance condition then reads

$$\omega - k_\varphi v_\varphi = l\Omega_2, \tag{3.24}$$

which may be interpreted as a Doppler-shifted bounce resonance by analogy with (2.38). On the other hand, if $|m\Omega_3| \gg |l\Omega_2|$, it may be instructive to view (3.23) as a bounce-modified drift resonance. Since the two interpretations are fully equivalent for any m and l, however, the connection with radial diffusion (Section III.3) is quite evident.

A similar connection may be drawn between the magnetic impulses of Section III.2 and an MHD wave propagating partly in the direction of ∇L_d and partly in the directions of $\hat{\mathbf{B}}$ and $\hat{\boldsymbol{\varphi}}$. The electric-field perturbation $\mathbf{e}$ for an MHD mode is normal to $\mathbf{k}$ and $\hat{\mathbf{B}}$ (in the cold-plasma approximation), and thus lies in the plane of $\hat{\boldsymbol{\varphi}}$ and ∇L_d. The $\hat{\boldsymbol{\varphi}}$ component of $\mathbf{e}$ leads to radial diffusion, the $\hat{\boldsymbol{\varphi}}$ component of $\mathbf{k}$ to drift resonance, and the $\hat{\mathbf{B}}$ component of $\mathbf{k}$ to bounce resonance. The condition imposed by (3.23) includes both bounce resonance and drift resonance. Either can be isolated by assigning $m=0$ or $l=0$, respectively.

III.5 Cyclotron Resonance

Because it leads to substantial pitch-angle diffusion, the Doppler-shifted $(k_\parallel v_\parallel)$ cyclotron resonance considered in Section II.5 is principally a loss mechanism for geomagnetically trapped particles. Cyclotron resonance is not known to be an important mechanism for *radial* diffusion in the radiation belts. A particle is perhaps displaced by one gyroradius in the course of diffusing by one radian in pitch angle. In the absence of shell splitting (see Sections I.7 and III.7), the resulting radial-diffusion coefficient is of order $\varepsilon^2 L^2 D_{xx}$. This is rather insignificant for radiation-belt ($|\varepsilon| \ll 1$; Section I.1) particles, since the root-mean-square displacement in L is only of order εL during the lifetime of a particle (in weak diffusion; see Section II.7). The most energetic radiation-belt ions (for which $|\varepsilon|$ is nearest to unity) tend to deposit their energy in the tenuous atmosphere without significant change of pitch angle.

The conditions under which radial diffusion might occur by virtue of cyclotron resonance are quite different from the conditions explored in Section II.5. Consider, for example, a wavelike electrostatic potential of the form

$$V_e(r,\theta,\varphi;t)=a\,E_m(L_d)\sin(m\varphi-\omega t+\psi_m) \qquad (3.25)$$

where $\omega/2\pi$ is of the order of a particle's gyrofrequency. A wave of this type may be generated by virtue of an unstable spatial gradient of $\bar{f}$, e.g., $\partial\bar{f}/\partial L_d<0$, with the conserved quantities held constant. Such an azimuthally propagating wave is called a *drift wave*, whether or not cyclotron resonance is involved.

The unperturbed motion of an equatorially mirroring particle may be represented by

$$\varphi=\Omega_3 t+\varphi_3+(pc/q\,B\,L_d a)\sin(\Omega_1 t+\varphi_1). \qquad (3.26)$$

The postulated drift wave does not alter the equatorial pitch angle $(\pi/2)$ of such a particle. The particle's interaction with the wave specified by (3.25) yields a φ_3-dependent drift in L $(\equiv L_d$ in a dipole field) given by

$$dL/dt=-(c/a)(L^3/B_0)m\,E_m(L)\sum_l J_l(-mv_\perp/\Omega_1 La)$$
$$\times\cos[(\omega-l\Omega_1-m\Omega_3)t-(\psi_m+l\varphi_1+m\varphi_3)]. \qquad (3.27)$$

The resonance condition $\omega=\omega_{lm}\equiv l\Omega_1+m\Omega_3$ leads to a radial-diffusion coefficient of the form

$$D_{LL}=2(c/2a\,B_0)^2\,L^6\sum_{lm}m^2\,J_l^2(mv_\perp/\Omega_1 La)\,\mathscr{E}_m(L,\omega_{lm}/2\pi), \qquad (3.28)$$

where $\mathscr{E}_m(L,\omega/2\pi)$ is the spectral density of all waves having the form of (3.25). Note that m represents an *azimuthal index*, not a mass, in (3.25)—(3.28). The leading factors in (3.22) and (3.28) differ only because each term in (3.21) is a superposition of two waves having the form of (3.25).

In the argument of the Bessel function J_l, the factor m/La plays the role of k_φ in (3.24) or of $k_\perp$ in (2.43). Thus, for azimuthal wavelengths comparable to a particle's gyroradius, a drift wave can resonate with the gyration of the particle in a manner that leads to radial diffusion. The pitch-angle of an equatorially mirroring particle is unaffected by this process, but the energy of a resonant particle changes in accordance with the relation

$$d(p^2)/dL=2q\,\mathbf{p}\cdot\mathbf{e}(dL/dt)^{-1}=2q\omega a^2\,B_0\gamma m_0/mc\,L^2 \qquad (3.29)$$

in a localized region where $dE_m/dL\approx0$. The ratio m/q is negative for a wave (m) propagating in the direction of the resonant particle's azimuthal drift. Thus, an outward diffusive flow of trapped particles (arising from an inward gradient of $\bar{f}$ with respect to L) leads to a transfer of particle energy to the wave[24]. The interaction evidently conserves

$$p^2 - 2(q/m)(a^2 B_0 m_0/c)\int(\gamma\,\omega/L^2)dL = \text{constant}, \qquad (3.30)$$

and so this is the quantity that must be held constant in evaluating $\partial\bar{f}/\partial L$. Although not known to play an essential role in radiation-belt dynamics, drift waves represent a potentially significant mechanism for extracting free energy from magnetospheric particle distributions by causing diffusion across field lines.

III.6 Bohm Diffusion

Electric Drift Velocity. In the absence of collisions and wave-particle interactions, the response of a charged particle to an electric field $\mathbf{E}_\perp$ imposed across $\mathbf{B}$ is the drift given by (1.53) or (3.06). For a simple derivation of this fact, consider that the transverse (to $\mathbf{B}$) electric field vanishes in a Lorentz frame moving at velocity $\mathbf{v}_0$ such that

$$c\,\mathbf{E}_\perp + \mathbf{v}_0 \times \mathbf{B} = 0. \qquad (3.31)$$

If $\mathbf{B}$ is uniform, a particle can only gyrate in this frame and execute translational motion along $\mathbf{B}$. The cross product between $\mathbf{B}$ and (3.31) then yields (1.53) or (3.06) as the velocity of the Lorentz transformation, *i.e.*, of the guiding-center motion across $\mathbf{B}$. If $\mathbf{B}$ is not uniform, then there are additional guiding-center forces equivalent to $q\mathbf{E}$. Replacement of $q\mathbf{E}$ in (1.53) or (3.06) by the sum of all forces $\mathbf{F}$ acting on a particle yields a drift velocity

$$\mathbf{v}_d = (c/qB^2)\mathbf{F} \times \mathbf{B}. \qquad (3.32)$$

The validity of (1.53) or (3.06) requires only that $v < c$. Guiding center forces requiring an average over gyration, *e.g.*, the forces $-(M/\gamma)\nabla B$ and $-(p_\parallel^2/m)(\partial\mathbf{B}/\partial s)$ leading to gradient and curvature drifts (see Section I.5), limit (3.32) to drift velocities much less than $\varepsilon\Omega_1 La$ in absolute value. Since the gradient-curvature drift velocity is in fact of order $\varepsilon^2\Omega_1 La$, this means only that the general validity of (3.12) is limited to $|\varepsilon| \ll 1$, as previously assumed.

Effective Collision Frequency. In causing diffusion with respect to energy, pitch-angle, and L value, wave-particle interactions have an effect quite analogous to that of interparticle collisions. For this reason it is often

---

[24]Drift waves can be destabilized under a variety of conditions [60]. The present calculation illustrates only one example.

convenient to think in terms of a effective collision frequency $1/\tau_\perp$ to which the various diffusion coefficients can be related, just as if interparticle collisions were the agent responsible for the diffusion. This equivalent collision frequency is said to produce anomalous transport, in the sense that the diffusion exceeds that which would result from Coulomb collisions acting alone. Thus, the quantity $1/\tau_\perp$ generally exceeds the Coulomb collision frequency.

The mean (phase-averaged) force exerted by collisions and wave-particle interactions can be represented by $-(m/\tau_\perp)\mathbf{v}_d$. If $\mathbf{B}$ is uniform, therefore, the net drift velocity resulting from the imposition of $\mathbf{E}_\perp$ across $\mathbf{B}$ is given by

$$\mathbf{v}_d = (c/B)\,\mathbf{E}_\perp \times \hat{\mathbf{B}} + (1/\Omega_1\tau_\perp)\mathbf{v}_d \times \hat{\mathbf{B}}$$
$$= \frac{(c\,E_\perp/B)(\Omega_1\tau_\perp)}{1+(\Omega_1\tau_\perp)^2}\left[\Omega_1\tau_\perp(\hat{\mathbf{E}}_\perp \times \hat{\mathbf{B}}) - \hat{\mathbf{E}}_\perp\right], \tag{3.33}$$

where $\Omega_1 = -q\,B/mc$. This result indicates a *Hall mobility*

$$\mu_H = (c/B)(\Omega_1\tau_\perp)^2\left[1+(\Omega_1\tau_\perp)^2\right]^{-1} \tag{3.34a}$$

in the direction of $\hat{\mathbf{E}}_\perp \times \hat{\mathbf{B}}$ and a *Pedersen mobility*

$$\mu_\perp = -(c/B)(\Omega_1\tau_\perp)\left[1+(\Omega_1\tau_\perp)^2\right]^{-1} \tag{3.34b}$$

in the direction of $\hat{\mathbf{E}}_\perp$. The Pedersen mobility approaches zero in the limit of no "collisions" ($\Omega_1^2\tau_\perp^2 \gg 1$) and approaches $q\tau_\perp/m$ in the limit of "collision" dominance ($\Omega_1^2\tau_\perp^2 \ll 1$). The maximum absolute value ($c/2B$) of $\mu_\perp$ is attained when $\Omega_1^2\tau_\perp^2 = 1$.

Diffusion Coefficient. The purpose of calculating the Pedersen mobility is ultimately to obtain the diffusion coefficient related to it, *i.e.*, the coefficient for the stochastic transport of particles *across* adiabatic drift shells. The connection between mobility and diffusion is given [61] by $D_\perp = (p_\perp^2/2qm)\mu_\perp$. Since L is a dimensionless variable scaled by the earth radius a, the quantity $D_\perp$ must be interpreted as $a^2 D_{LL}$. It follows that

$$D_{LL} = (p_\perp/ma)^2(\tau_\perp/2)\left[1+(\Omega_1\tau_\perp)^2\right]^{-1}. \tag{3.35}$$

If $\tau_\perp$ is now considered an adjustable parameter, the magnitude of D_{LL} can be maximized by setting $\tau_\perp = |\Omega_1|^{-1}$. In other words, there exists an upper bound, given by

$$D_{LL}^* = (p_\perp/2ma\,\Omega_1)^2|\Omega_1|, \tag{3.36}$$

on the coefficient of radial diffusion. No adjustment of $\tau_\perp$ can produce a value of D_{LL} larger than D_{LL}^*. A process in which $D_{LL} \sim D_{LL}^*$ is characterized as *Bohm diffusion* [62]. It represents the most expedient means

available to a hot plasma for erasing an unstable spatial gradient (*cf.*
Section III.5) in the distribution function, and in this sense is analogous
to strong pitch-angle diffusion (Section II.7), which has the same property
relative to unstable gradients of $\bar{f}$ in momentum space.

There is, however, no reason why Bohm diffusion must cause strong
pitch-angle diffusion. As in Section III.5, the "collisions" could easily
act preferentially in the direction normal to **B**, an option not available
to interparticle collisions. Thus, the anomalous Ohmic mobility
$\mu_{\parallel} \equiv (\mathbf{v} \cdot \hat{\mathbf{B}})/(\mathbf{E} \cdot \hat{\mathbf{B}})$ is given by $q\tau_{\parallel}/m$, where $\tau_{\parallel}$ may be entirely different in
magnitude from $\tau_{\perp}$ in (3.34). In the event that $\tau_{\parallel} \gg \tau_{\perp}$, there may be
very few particles scattered into the loss cone in the course of Bohm
diffusion. Conversely, strong diffusion requires only that $\Omega_2 \tau_{\perp} \ll 1$ and
$\Omega_2 \tau_{\parallel} \ll 1$. These conditions do not necessarily imply $|\Omega_1| \tau_{\perp} \sim 1$, as
required for Bohm diffusion.

An examination of (3.36) indicates that $D_{LL}^* \sim L^2 |\Omega_3|$. No radiation-
belt observations are known to require nearly this large a value of
D_{LL}, but the storm-time ring current occasionally appears to exhibit
Bohm diffusion in the vicinity of the plasmapause. The plasmasphere
tends to destabilize the ring current against electromagnetic ion-cyclotron
turbulence (see Section II.6) by drastically reducing the minimum
resonant energy given by (2.69b). Bohm diffusion is sometimes invoked
[63], in addition to the adiabatic gradient-curvature drift, as a means
of transporting ring-current protons into the plasmasphere from the
exterior region in which N_p is very small ($\sim 0.1 \, \text{cm}^{-3}$ during a magnetic
storm). Even in the presence of strong pitch-angle diffusion, which
the resulting ion-cyclotron turbulence causes, the Bohm diffusion coefficient would transport ring-current protons ($\varepsilon \sim 10^{-3}$) a root-mean-square
distance $\sim 0.5a$ relative to the plasmapause during the lifetime $1/\lambda$
given by (2.77).

III.7 Shell Splitting

As described in Section I.7, drift-shell splitting is a purely adiabatic
phenomenon that violates none of the invariants. Radial diffusion, by
definition, violates the third invariant. Pitch-angle diffusion violates
either or both of the first two invariants, usually both. In a symmetrical
magnetosphere, the incidentally associated radial diffusion coefficient
$D_{LL}(\sim \varepsilon^2 L^2 D_{xx})$ would be too small to be of significance for radiation-belt
particles ($\varepsilon^2 \ll 1$). In the presence of azimuthal asymmetry and shell
splitting, however, pitch-angle diffusion *automatically* produces an additional violation of the third invariant. The shell-tracing results obtained

in Section I.7 permit this effect to be evaluated for arbitrary values of the equatorial pitch angle.

The basic equation governing the process under consideration is (*cf.* Section II.2)

$$D_{LL} = \langle (\partial L/\partial x)^2 D_{xx} \rangle = \langle (x/y)^2 (\partial L/\partial y)^2 D_{xx} \rangle. \tag{3.37}$$

Thus, if the values of $L (\equiv 2\pi a^2 B_0 |\Phi|^{-1})$, among identical particles having mirror points on a common field line, vary with equatorial pitch angle, then pitch-angle diffusion of these particles on this field line automatically produces diffusion with respect to L [5]. The partial derivatives are evaluated by holding constant the quantity conserved by D_{xx}, typically the particle energy or first invariant. The drift average denoted by the angle brackets necessarily yields a positive-definite D_{LL} [65].

External Multipoles. In the case of magnetic shell splitting, as summarized by (3.11), pitch-angle diffusion leaves L_d and φ invariant at the scattering site, and so the quantity $\partial L/\partial y$ is given by

$$\partial L/\partial y = -(B_2/252 B_0)(a/b)^4 L_d^5 [D(y)]^{-2}$$
$$\times [Q'(y)D(y) - D'(y)Q(y)]. \tag{3.38}$$

If pitch-angle diffusion is distributed uniformly with respect to longitude, *i. e.*, if D_{xx} is independent of φ, then to lowest order in $\varepsilon_2 \equiv (B_2/B_0)(L_d a/b)^4$ it follows that

$$D_{LL} = (x^2/2 y^2)(B_2/252 B_0)^2 (a/b)^8 L^{10} [D(y)]^{-4}$$
$$\times [Q'(y)D(y) - D'(y)Q(y)]^2 D_{xx}. \tag{3.39}$$

With the aid of (1.36) and (1.79), the function $(x^2/98 y^2)[6D(y)]^{-4}$ $\times [Q'(y)D(y) - D'(y)Q(y)]^2$, which expresses the pitch-angle dependence of D_{LL}/D_{xx}, has been plotted in Fig. 28 [66]. This function reduces to $25 x^2/18$ in the limit $x^2 \equiv 1 - y^2 \ll 1$, in which case (3.39) reduces to the expression

$$D_{LL} \approx (x^2/18)(5 B_2/B_0)^2 (a/b)^8 L^{10} D_{xx}$$
$$\approx 0.61 x^2 (a/b)^8 L^{10} D_{xx}. \tag{3.40}$$

As an upper bound on radial diffusion induced by magnetic shell splitting, this expression remains valid for $|x| \lesssim 0.9997$; it fails only deep within the loss cone. As an approximate expression for D_{LL}, equation (3.40) remains valid within a factor of two only for $|x| \lesssim 0.6$, while (3.39) is correct to within a few percent for all x when evaluated via (1.36) and (1.79).

The complete Jacobian entering (2.12), when pitch-angle diffusion violates Φ, depends upon the nature of the quantity conserved in the

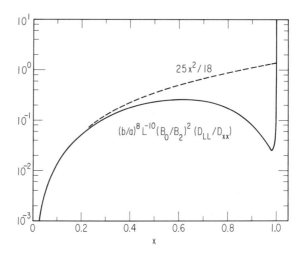

Fig. 28. Relation between D_{LL} and D_{xx} for shell splitting caused by noon-midnight magnetic asymmetry [66], as given by harmonic-bounce approximation (dashed curve) and by improved approximation (solid curve) based on (1.36) and (1.79).

process. If the conserved quantity is particle energy, then the relevant Jacobian is

$$G(M,J,|\Phi|;E,x,L)= -8\pi\gamma p L^2 a^3 x T(y), \qquad (3.41\,a)$$

as deduced from (2.14) and (1.37). If, as in the case of bounce resonance, the conserved quantity is M, then the relevant Jacobian is

$$G(M,J,|\Phi|;M,x,L)= -8\pi B_0 a^3 (p/y^2 L)x T(y), \qquad (3.41\,b)$$

which follows from (2.27), (1.37), and the fact that $(\partial w/\partial x)_{M,L} = 2MB_0 x/L^3 y^4$ [cf. (2.33) and (2.34)].

In (3.41a), the distribution function $\bar{f}$ is considered to depend on E, x, and L. Since E is conserved by the process, the distribution function satisfies [65]

$$\frac{\partial \bar{f}}{\partial t}=\frac{1}{L^2}\frac{\partial}{\partial L}\left[L^2 D_{LL}\frac{\partial \bar{f}}{\partial L}\right]_x+\frac{1}{x T(y)}\frac{\partial}{\partial x}\left[x T(y)D_{xx}\frac{\partial \bar{f}}{\partial x}\right]_L, \quad (3.42)$$

where D_{LL} is given by (3.39). This equation contrasts strikingly with (3.01), which applies to processes that conserve M and J.

Electric Shell Splitting. In the case of electric shell splitting, caused by superposition of (1.52) upon the dipole field (1.16), the relation between L and y at constant L_d and φ is expressed by (3.15), provided that

$W \gg |q V_e|$ around the entire drift shell. The connection between D_{LL} and a φ-independent D_{xx} is then given by

$$D_{LL} \approx (q E_c a L^2/W)(m_0 c^2 + W)^2 (2 m_0 c^2 + W)^{-2} [6 D(y)]^{-4}$$
$$\times [Y'(y) T(y) - T'(y) Y(y)]^2 (x^2/2 y^2) D_{xx} \qquad (3.43)$$

for pitch-angle diffusion at constant particle energy W. With the aid of (1.28), (1.31), and (1.36), the function $(x^2/8 y^2)[6D(y)]^{-4}[Y'(y)T(y) - T'(y)Y(y)]^2$ has been plotted in Fig. 29. This function indicates the pitch-angle dependence of D_{LL}/D_{xx} in the presence of electric shell splitting, and approaches $x^2/162$ in the limit $x^2 \ll 1$. The nonrelativistic limit ($W \ll m_0 c^2$) of (3.43) therefore reads

$$D_{LL} \approx (x^2/162)(q E_c a L^2/W)^2 D_{xx} \qquad (3.44)$$

for $x^2 \ll 1$, and represents a serious (factor-of-two) underestimate for D_{LL} only if $|x| \gtrsim 0.6$.

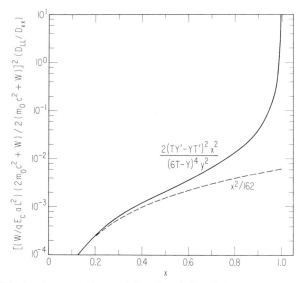

Fig. 29. Relation between D_{LL} and D_{xx} for shell splitting caused by dawn-dusk asymmetry of electrostatic potential [66], as given by harmonic-bounce approximation (dashed curve) and by improved approximation (solid curve) based on (1.28), (1.31), and (1.36).

A comparison between (3.40) and (3.44), assuming $b = 10 a$ and $E_c a = 4 \text{kV}$, suggests that magnetic and electric shell-splitting effects are comparable at $M/y^2 \sim 7 \text{ MeV/gauss}$, *i.e.*, at first invariants typical of the ring-current particles. In the true *radiation* belts, magnetic shell splitting effects exceed those of electric shell splitting[25].

At particle energies below those typical of the ring current, it is necessary to reconsider the shell-splitting problem in terms of (1.65). Beyond the plasmapause, such drift shells do not close within the magnetosphere, and the corresponding third invariants are undefined (*cf.* Fig. 12, Section I.6). Within the plasmasphere, all shell splitting disappears in the cold-plasma limit, since "zero-energy" particles drift on field-aligned surfaces of constant electrostatic potential.

Internal Multipoles. At very low L values, certain internal geomagnetic multipoles associated with true field anomalies, may cause significant shell splitting among inner-belt particles [67]. If electric fields are negligible, the existence of magnetic shell splitting in general can be demonstrated (*cf.* Section I.7) by showing that $(\partial^2 B/\partial s^2)_e$ varies with φ around a path of constant B_e on the equatorial $(\partial B/\partial s = 0)$ surface. This criterion follows from (1.26) and (1.32a), in the sense that the drift shell (which conserves M and J) must depart from the constant-B_e trajectory for $x \neq 0$ if Ω_2 varies with φ; to lowest order in x, the bounce frequency is given by $\Omega_2^2 = (M/\gamma m)(\partial^2 B/\partial s^2)_e$.

In a dipole field, the value of $(\partial^2 B/\partial s^2)_e$ is given by $9 B_0/a^2 L_m^5$, where $L_m^3 \equiv (B_0/B_e)$ at $y = 1$ [*cf.* (1.38)]. It proves convenient to display the azimuthal variation of $(\partial^2 B/\partial s^2)_e$ at constant B_e and L_m by plotting $L_m^2[(L_m^5 a^2/9 B_0)(\partial^2 B/\partial s^2)_e - 1]$ against geomagnetic longitude. This is done[26] in Fig. 30 for $L_m = 1,2$, and 7. The sinusoidal asymptote, approximated by the curve $L_m = \infty$, results from an internal octupole. The octupole corresponds to $n = -5$ in (1.46) and the dipole to $n = -3$, hence the factor L_m^2 in Fig. 30. Higher multipoles produce the broad South African anomaly (longitude $30°$—$120°$) and the narrow South American anomaly (longitude $0°$—$30°$). The latter disappears between $L_m = 1$ and $L_m = 2$. The fact that the dipole is off center by $\sim 0.07 a$, toward longitude $217°$, necessarily contributes nothing to shell splitting as this is not a *field* asymmetry. Components of the geomagnetic quadrupole that survive the transformation to offset-dipole coordinates can only warp the equatorial $(\partial B/\partial s = 0)$ surface as a lowest-order effect. Their second-order (shell-splitting) effects are not discernible in Fig. 30 [67].

---

[25]However, the demarcation between ring-current and radiation-belt parameters is somewhat arbitrary (Fig. 13, Section I.7).

[26]The shell $L_m = 1$ is unphysical in the sense that it intersects the earth's surface (see Section II.7).

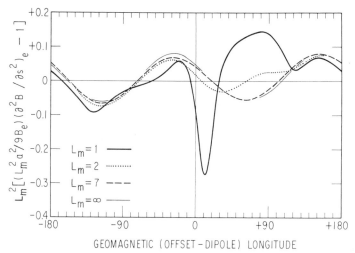

GEOMAGNETIC (OFFSET–DIPOLE) LONGITUDE

Fig. 30. Normalized shell-splitting function associated with internal geomagnetic multipoles, shown for selected contours of constant B on the equatorial surface [67].

True field anomalies (including the octupole) that significantly split drift shells thereby subject the inner radiation belt to radial diffusion coincident with pitch-angle scattering. A lower bound on the resulting D_{LL} is provided by the inequality

$$D_{LL} \gtrsim (L_m x/3)^2 (L_m^5 a^2/9 B_0)^2$$
$$\times \text{Min} \left\langle [(\partial^2 B/\partial s^2)_0 - (\partial^2 B/\partial s^2)_e]^2 D_{xx} \right\rangle \qquad (3.45)$$

where the angle brackets denote an equatorial drift average, which must be minimized with respect to some reference longitude φ_0 at which $(\partial^2 B/\partial s^2)_e \equiv (\partial^2 B/\partial s^2)_0$. The minimizing operation (Min) assures that (3.45) is a lower bound on the radial diffusion coefficient, regardless of the reference longitude that ultimately proves suitable for defining L [cf. (3.10)]. If the pitch-angle scattering is principally atmospheric (e.g., at $L_m \lesssim 1.25$ for inner-belt electrons), then the magnitude of D_{xx} is strongly φ-dependent, with a peak near geomagnetic longitude $20°$ (cf. Section II.2).

III.8 Diffusion in More Than One Mode

For each diffusion mechanism considered in Chapters II and III, the diffusion tensor D_{ij} (see Section II.1) can be diagonalized by a proper choice of variables, i.e., by transforming from the coordinates (M, J, Φ)

to an equivalent set of functionally independent variables. Mixed partial derivatives in (2.12) are thus eliminated. Vanishing eigenvalues (diagonal elements, $i=j$) of the transformed diffusion tensor $\tilde{D}_{ij}$ correspond to conservation laws of the diffusion mechanism [68]. For example, *pure* pitch-angle diffusion, *i.e.*, diffusion at constant particle energy, corresponds to $D_{EE}=0$ and (in the absence of shell splitting) $D_{LL}=0$. Pure third-invariant diffusion (Sections III.1—III.3) has the property that $D_{MM}=D_{JJ}=0$. A summary of various diffusion mechanisms, their conservation laws, and the Jacobians of their respective diagonalizing transformations is given in Table 8.

Table 8. Diffusion Variables and Associated Jacobians

Interaction	Invariants	Relevant Jacobian
Elastic Collisions (without recoil)	$E, (\Phi)$	$\|G(M, J, \Phi; E, x, L)\|$ $=8\pi\gamma p L^2 a^3 x T(y)$
Cyclotron Resonance	$(E), (\Phi)$	$\|G(M, J, \Phi; E, x, L)\|$ $=8\pi\gamma p L^2 a^3 x T(y)$
Bounce Resonance	$M, (\Phi)$	$\|G(M, J, \Phi; M, x, L)\|$ $=8\pi(a/y)^3(2m_0 B_0^3 M/L^5)^{1/2} x T(y)$
Drift Resonance	M, K	$\|G(M, J, \Phi; M, K, L)\|$ $=(8m_0 M)^{1/2}(2\pi B_0 a^2/L^2)$
Bimodal Diffusion	(ζ)	$\|G(M, J, \Phi; \zeta, x, L)\|$ $=8\pi a^3(2m_0 B_0^3 \zeta/L^5)^{1/2} x T(y)$

(Parenthesized "invariant" quantities are either approximately or conditionally conserved).

No special difficulty of concept arises when two or more diffusion mechanisms act simultaneously. If the concurrent processes satisfy the same conservation laws, then a single transformation of variables will suffice to make the diffusion tensor diagonal. If not, *i.e.*, if the conservation laws for kinematical variables are not common to the various diffusion mechanisms acting concurrently, then the problem is said to involve more than one *mode* of diffusion. In this case, the diffusion equation is at least two-dimensional with respect to the kinematical variables. This property presents no special difficulty, since two-dimensional diffusion equations, *e.g.*, (3.42), have already appeared in the context of unimodal diffusion. In constructing a bimodal diffusion equation, however, it is essential to evaluate the partial derivatives in accordance with the conservation laws of the respective modes. For example, if radial diffusion at constant M and J (coefficient D_{LL}) is superimposed upon pitch-angle diffusion at constant E (coefficient D_{xx}) in the presence of magnetic shell splitting, the equation governing this bimodal process is

$$\frac{\partial \bar{f}}{\partial t} = L^2 \frac{\partial}{\partial L} \left[\frac{1}{L^2} D_{LL} \frac{\partial \bar{f}}{\partial L} \right]_{M,J} + \frac{1}{x\,T(y)} \frac{\partial}{\partial x} \left[x\,T(y)\,D_{xx} \frac{\partial \bar{f}}{\partial x} \right]_{E,L}$$
$$+ \frac{1}{L^2} \frac{\partial}{\partial L} \left[\frac{x^2 L^{12} [Q'(y)D(y) - D'(y)Q(y)]^2 D_{xx}}{2(252\,B_0/B_2)^2 (b/a)^8 [D(y)]^4 \, y^2} \frac{\partial \bar{f}}{\partial L} \right]_{E,x},$$
\hfill (3.46)

a result obtained by consolidating (3.01), (3.39), and (3.42).

The right-hand side of (3.46) has the form of minus the "divergence" of a diffusion current for each mode (*cf.* Sections II.1 and II.2). The radial (trans-L) component of the diffusion current has the form $-D_{LL}(\partial \bar{f}/\partial L)_{M,J}$ for the sudden-impulse mode and the form $-\langle (x/y)^2 (\partial L/\partial y)^2 D_{xx} \rangle (\partial \bar{f}/\partial L)_{E,x}$ for the shell-splitting mode [*cf.* (3.37), Section III.7]. For outer-belt electrons at $L \gtrsim 5$, it is interesting that $(\partial \bar{f}/\partial L)_{M,J}$ is typically positive, while $(\partial \bar{f}/\partial L)_{E,x}$ is typically negative (see Fig. 1 and Section IV.6). The diffusion current across L thus consists of an *inward* part conserving M and J, which tends to energize the diffusing particles, and an *outward* part conserving E. The net result is that, for particles diffusing "bimodally" from an external source into the outer belt, the gain in energy typically *exceeds* that predicted on the basis of constant M and J [69] (see Section III.1).

Even if shell-splitting effects are neglected, *e.g.*, by taking $B_2 = 0$, the diffusion equation (3.46) is two-dimensional in the sense that no overall conservation law relates x and L. Thus, an individual particle from the distribution $\bar{f}(E, x, L; t)$ may random-walk a complete cycle in x and L, as illustrated in Fig. 31. In the absence of shell splitting,

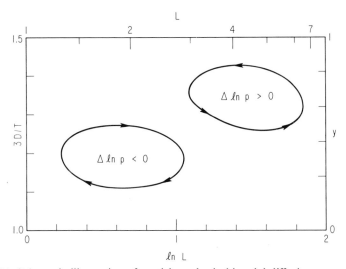

Fig. 31. Schematic illustration of particle cycles in bimodal diffusion.

the radial diffusion in (3.46) occurs at constant M and J. According to (1.34 b), the variation of particle energy with L is governed by the relationship

$$(\partial \ln p / \partial \ln L)_{M,J} = -3[D(y)/T(y)]. \qquad (3.47)$$

It follows that a clockwise cycle in Fig. 31 (inward diffusion at generally smaller y than outward diffusion) represents a net loss in particle energy, while a counter-clockwise cycle causes a particle to gain energy. In this context, bimodal diffusion acts as a "thermalization" mechanism, whereby an initially narrow energy spectrum of particles can become distributed to both higher and lower energies than pure conservation of M and J would allow [69].

Reduced Diffusion Equations. For many problems involving radiation-belt diffusion, it is considered appropriate to simplify (3.46) by means of approximations that reduce the diffusion equation to one spatial dimension. Simplifying approximations of this type are often indicated when the observational data are not sufficiently complete to impose meaningful boundary conditions on (3.46). In many cases the observations cover too limited a range of parameter space to make full use of (3.46). Reduction of the diffusion equation to one dimension, however justified, does require that bimodal cycles of the type illustrated in Fig. 31 be neglected. This is part of the cost of analytical simplification.

A naive means of reducing (3.46) is to neglect shell-splitting effects and to replace the pitch-angle diffusion term by a simple loss term of the form $-\bar{f}/\tau$. In this approximation [70] the diffusion equation reads [cf. (2.09)]

$$\frac{\partial \bar{f}}{\partial t} = L^2 \frac{\partial}{\partial L}\left[\frac{1}{L^2}\bar{D}_{LL}\frac{\partial \bar{f}}{\partial L}\right]_M - \frac{\bar{f}}{\tau} \qquad (3.48)$$

and applies to $\bar{f}(M,L;t)$ at $J=0$. The pitch angles of particles having in common their values of M/y^2 and L are mixed thoroughly on a time scale $\sim \tau/5$ (see Section II.7). The representation of pitch-angle diffusion as a simple loss term, as in (3.48), essentially requires that $5\bar{f}/\tau$ greatly exceed $\partial \bar{f}/\partial t$ in absolute value[27]. The diffusion coefficient $\bar{D}_{LL}$ is then interpreted as an average over particles sharing the same values of M/y^2 and L, respectively.

A more sophisticated view of the reduction described in the paragraph above is that a new variable $\zeta \equiv M/y^2$ has been introduced, and that ζ is *approximately* conserved by both D_{LL} and D_{xx} [71]. From this

---

[27]This requirement is often overlooked in the interest of expedience.

viewpoint, the form of (3.48) should be governed by the Jacobian [5]

$$G(M, J, |\Phi|; \zeta, x, L) = -(8\pi a^3 B_0/L^{5/2}) x\, T(y)(2m_0 B_0 \zeta)^{1/2}, \quad (3.49)$$

which has been included in Table 8. With this Jacobian, the reduced (to one dimension) diffusion equation evidently has the form

$$\frac{\partial \bar{f}}{\partial t} = L^{5/2} \frac{\partial}{\partial L}\left[L^{-5/2} \bar{D}_{LL} \frac{\partial \bar{f}}{\partial L}\right]_\zeta - \frac{\bar{f}}{\tau}. \quad (3.50)$$

The practical discrepancy between (3.48) and (3.50) is slight, amounting only to a square root of L in the metric. Since D_{LL} typically varies as L^{10} (see Sections III.2 and III.3) for radiation-belt particles, it is difficult to imagine that seriously different geophysical predictions might emerge from (3.48) and (3.50), although (3.50) is perhaps preferable in terms of self-consistency.

In either representation the transport coefficients may certainly vary with L, and perhaps also with ζ ($=M$ at $J=0$) and/or time. Since $\bar{D}_{LL}$ and τ arise from operations on the entire pitch-angle distribution, it would be meaningless to give either a dependence on x or y. This degree of freedom has been sacrificed in reducing (3.46) to one dimension. The conservation of ζ is clearly an idealization that breaks down for $x \sim 1$, but the presence of a loss cone (see Section II.7) assures that $\bar{f}$ is small there[28]. Thus, the effective radial-diffusion coefficient $\bar{D}_{LL}$ is heavily weighted by the behavior of particles for which $x^2 \ll 1$, i.e., for which radial diffusion at constant M and J very nearly conserves ζ.

If the time scale for pitch-angle mixing ($\sim \tau/5$) is comparable to that for radial diffusion, then a simplified equation such as (3.50), which assigns to $\bar{f}$ the lowest mode of pitch-angle diffusion (see Section II.7), cannot apply unless D_{LL} is substantially independent of x (cf. Sections III.2 and III.3). Thus, radial diffusion caused by electrostatic impulses may lend itself to analysis via (3.50), but that caused by magnetic impulses will ordinarily bias $\bar{f}$ toward higher modes of pitch-angle diffusion. In this case a more general treatment is required.

If the need to circumvent (3.46) is compelling, it may be possible to expand $\bar{f}(\zeta, x, L; t)$ in pitch-angle eigenfunctions $g_n(x)$ that are even in x (even parity required because of homogeneity over bounce phase). An expansion [71] of the form

$$\bar{f}(\zeta, x, L; t) = \sum_n a_n(\zeta, L; t) g_n(x) \quad (3.51)$$

---

[28]The constant-ζ approximation means that particles diffuse radially at constant y, in weak violation of (1.34a), (3.02), and Fig. 24.

with the boundary condition $g_n(x_c)=0$ is justified if x_c is independent of L, and the L dependence of D_{xx} is factorable, $i.e.$, if D_{xx} is the product of a function of L, ζ, and t times a function of x. These conditions on x_c and D_{xx} are probably well satisfied in the outer zone. It is convenient to assume further that D_{xx} and D_{LL} are time-independent. In this case the approximate diffusion equation [$cf.$ (3.50), (3.49), and (3.46), with $B_2=0$ ($i.e.$, without shell splitting)]

$$\frac{\partial \bar{f}}{\partial t} = L^{5/2} \frac{\partial}{\partial L}\left[L^{-5/2} D_{LL} \frac{\partial \bar{f}}{\partial L}\right]_x + \frac{1}{x\,T(y)} \frac{\partial}{\partial x}\left[x\,T(y) D_{xx} \frac{\partial \bar{f}}{\partial x}\right]_L \qquad (3.52\,\text{a})$$

can be simplified by virtue of the eigenvalue property

$$\frac{1}{x\,T(y)} \frac{\partial}{\partial x}\left[x\,T(y) D_{xx} g_n(x)\right]_L = -\lambda_n(\zeta, L)g_n(x), \qquad (3.52\,\text{b})$$

where $\lambda_n(\zeta,L)$ is the decay rate characteristic of the pitch-angle eigenmode g_n ($cf.$ Section II.7).

The normalized eigenfunctions corresponding to distinct eigenvalues λ_n are orthogonal in the sense that

$$\int_0^{x_c} x\,T(y)g_n(x)g_m(x)dx = \delta_{nm}. \qquad (3.53)$$

Application of (3.52) to (3.51) therefore implies that

$$\frac{\partial a_m}{\partial t} = L^{5/2} \frac{\partial}{\partial L}\left[L^{-5/2} \sum_n \bar{D}_{LL}^{mn} \frac{\partial a_n}{\partial L}\right]_\zeta - \lambda_m a_m, \qquad (3.54\,\text{a})$$

where

$$\bar{D}_{LL}^{mn} \equiv \int_0^{x_c} x\,T(y) D_{LL} g_m(x)g_n(x)dx. \qquad (3.54\,\text{b})$$

If D_{LL} is independent of x, as is approximately true in radial diffusion caused by electrostatic impulses (see Section III.3), then the matrix $\bar{D}_{LL}^{mn}$ is diagonal in the sense that $\bar{D}_{LL}^{mn}=D_{LL}\delta_{mn}$. In this case the functions $a_m(\zeta,L;t)$ and $a_n(\zeta,L;t)$ in (3.54a) are decoupled for $m \neq n$ and diffuse separately with respect to L [71]. If $\bar{f}(\zeta,x,L;t)$ is initially in its lowest pitch-angle eigenmode $g_0(x)$, therefore, it will continue in this eigenmode and diffuse according to (3.50) as time goes on. On the other hand, off-diagonal elements of $\bar{D}_{LL}^{mn}$, which are obviously substantial in radial diffusion caused by $magnetic$ impulses (see Section III.2), serve to couple distinct pitch-angle eigenmodes and thereby "excite" modes not present in the initial configuration of $\bar{f}(\zeta,x,L;t)$.

Inner-Zone Protons. For particles that do *not* undergo significant pitch-angle diffusion, the fundamental radial-diffusion equation is (3.01). Very energetic ($E \gtrsim 100$ MeV) inner-zone protons are believed to be in this category. The principal source for these particles is known as CRAND (see Section III.1): cosmic rays incident on the upper atmosphere eject high-energy neutrons that beta-decay with a mean life τ_n ($\sim 10^3$ sec) in their own rest frame. At low latitudes the *vertical* flux J_n of these "albedo" neutrons is believed to be given [72] by

$$J_n \approx 0.044 \, (E/1 \, \text{MeV})^{-1.86} \, \text{cm}^{-2} \, \text{sec}^{-1} \, \text{MeV}^{-1} \qquad (3.55)$$

at the top of the atmosphere ($r = a + h$, *cf.* Sections II.2 and II.7). The presence of these decaying neutrons[29] requires that a proton *source term* [38]

$$S \approx (J_n / 2\pi\gamma\tau_n p^2)\chi \qquad (3.56a)$$

be added to the right-hand side of (3.01). The geometric injection coefficient χ for equatorially mirroring protons is estimated by the expression [73]

$$\chi \approx (2/\pi)\sin^{-1}[(a+h)/La]. \qquad (3.56b)$$

The arcsine represents the half angle subtended by the earth's atmosphere at the site of proton injection (neutron decay) in a model centered-dipole field[30].

The inner-zone protons injected by CRAND lose energy to free and bound ionospheric electrons [*cf.* (2.04)] but gain energy from the secular decrease of B_0 [*cf.* (2.05)]. Both processes leave the equatorial pitch angle invariant. The energy gain is an adiabatic effect, and so is automatically included if the problem is posed in the invariant coordinates M, J, and Φ, *i. e.*, in the form that reduces to

$$\frac{\partial \bar{f}}{\partial t} = S + \frac{\partial}{\partial \Phi}\left[D_{\Phi\Phi}\frac{\partial \bar{f}}{\partial \Phi}\right]_{M,\,K} + \frac{(4\pi q^4/m_e)}{(2MB_m^3/m_p)^{1/2}}\left[\frac{\partial(C\bar{f})}{\partial M}\right]_{K,\,\Phi}, \qquad (3.57a)$$

---

[29]The mean free path of a 100-MeV neutron before beta decay is of the order of one astronomical unit. Decay within the magnetosphere therefore does not significantly deplete the flux of cosmic-ray-albedo neutrons.

[30]For this derivation of S, it is assumed that the neutron flux is isotropic at the top of the atmosphere, so that the omnidirectional neutron flux $J_{2\pi}$ is twice the vertical flux J_n. The unidirectional neutron flux above the atmosphere remains $(1/2\pi)J_{2\pi}$, by Liouville's theorem (Section I.3), for gyrophase angles compatible with ejection from the atmosphere. The result is a gyrophase-averaged proton source $(d\bar{J}_\perp/dt)_n = (1/2\pi\gamma\tau_n)(\chi/2)J_{2\pi}$, *i.e.*, a source for $\bar{f}(=\bar{J}_\perp/p^2)$ given by (3.56).

where [cf. Section II.2]

$$C = \bar{N}_e[\gamma^2 - 1 - \gamma^2 \ln(\lambda_D m_e v/\hbar)]$$
$$+ \sum_i \bar{N}_i Z_i \{\gamma^2 - 1 - \gamma^2 \ln[2 m_e c^2(\gamma^2 - 1)/I_i]\} . \qquad (3.57\,b)$$

The mirror field B_m is given in terms of the invariant coordinate Φ by $B_m = (1/8\pi^3 a^6 y^2 B_0^2)|\Phi|^3$, and thus contains an explicit time dependence (that of B_0^{-2}). Expressed as functions of K^2 ($\equiv J^2/8 m_0 M$) and Φ, the drift-averaged atmospheric densities $\bar{N}_j$ also vary with time. The drift shell corresponding to given values of K and Φ not only contracts temporally (since $\dot{B}_0/B_0 < 0$), but also moves laterally relative to the earth so as to remain concentric with the dipole axis[31] (apart from the effects of magnetic anomalies, cf. Section III.7). A growing dipole-offset distance imparts an additional increase to $\bar{N}_j$ with time for atmospheric constituents whose densities decrease with altitude.

The secular variation[32] of B_0 on a time scale $\sim 2000\,\mathrm{yr}$ prevents (3.57) from having a steady-state ($\partial\bar{f}/\partial t = 0$) solution with which the inner proton belt can be identified. Thus, the present state of protons in the inner zone is the result of a long and continuing process of evolution. According to Fig. 14 (Section II.2) protons presently trapped in the inner zone may well have resided there for the past thousand years or more. An integration of (3.57) over this geomagnetic history may be fraught with uncertainty, in view of the available observations. Such a treatment appears to be necessary, however.

In much of the inner zone, the secular decrease of B_0 energizes trapped protons more efficiently than does inward radial diffusion at constant M and J. Typical time scales for the latter process at $J = 0$ have been indicated by broken lines in Fig. 14 (Section II.2). For this purpose, the diffusion "current" $-D_{LL}(\partial\bar{f}/\partial L)_{M,J}$ identified following (3.46) has been utilized to construct an effective "velocity"

$$\dot{L} = -D_{LL}(\partial \ln\bar{f}/\partial L)_{M,J} . \qquad (3.58)$$

Insertion of $10/L$ as a likely upper bound [38, 39] for $(\partial\ln\bar{f}/\partial L)_{M,J}$ leads to the estimate [cf. (2.05)] that

$$\frac{1}{E}\frac{dE}{dt} = \frac{\dot{L}}{B_e}\left[\frac{\gamma+1}{2\gamma}\right]\frac{\partial B_e}{\partial L} \lesssim \frac{30}{L^2}\left[\frac{\gamma+1}{2\gamma}\right]D_{LL} \qquad (3.59)$$

---

[31]At present the distance between the dipole axis and the geocenter is growing at a rate $\sim 2\,\mathrm{km/yr}$.

[32]Other axially symmetric internal multipoles (2^n) of odd-n order (e.g., octupole) may also contribute a significant secular variation having similar consequences [39].

if secular variation is neglected. The diffusion time scales shown in Fig. 14 (Section II.2) are obtained by inverting the right-hand side of (3.59) for representative values of D_{LL} at constant M and J (cf. Chapter V). These diffusion time scales are generally comparable to the secular and atmospheric time scales for e-folding the kinetic energy of an inner-belt proton having $M \sim 1\,\text{GeV/gauss}$.

Other Diffusion Velocities. Since radial diffusion is a macroscopically random (rather than deterministic) process, it may be possible to identify "velocities" other than (3.58) by following the temporal evolution of $\bar{f}$ in its various aspects. For example, the expansion of (3.01) as

$$\frac{\partial \bar{f}}{\partial t} - \frac{\partial D_{LL}}{\partial L} \left[\frac{\partial \bar{f}}{\partial L} \right]_{M,J} = L^2 D_{LL} \frac{\partial}{\partial L} \left[\frac{1}{L^2} \frac{\partial \bar{f}}{\partial L} \right]_{M,J} \qquad (3.60)$$

suggests the inward motion (at "velocity" $-\partial D_{LL}/\partial L$) of a diffusing profile of $\bar{f}$, viewed at an "inflection point" where $\partial^2 \bar{f}/\partial (L^3)^2 = 0$. Alternatively, if the distribution function has a symmetrical "crest" shell at which $(\partial \bar{f}/\partial L)_{M,J} = (\partial^3 \bar{f}/\partial L^3)_{M,J} = 0$, this "crest" can easily be shown to move at "velocity"

$$\dot{L}_c = -2(\partial D_{LL}/\partial L) + (2/L)D_{LL} \qquad (3.61)$$

when observed at fixed M and J. Finally, if (3.01) is recast as a Fokker-Planck equation [cf. (2.03)] of the form

$$\frac{\partial \bar{f}}{\partial t} = -L^2 \frac{\partial}{\partial L} \left[\frac{\bar{f}}{L^2} \frac{\partial D_{LL}}{\partial L} \right]_{M,J} + L^2 \frac{\partial}{\partial L} \left[\frac{1}{L^2} \frac{\partial (\bar{f} D_{LL})}{\partial L} \right]_{M,J}, \qquad (3.62)$$

the "velocity" $\partial D_{LL}/\partial L$ is seen to represent a mean displacement in L per unit time for the typical particle contained in $\bar{f}(M,J,L;t)$. Of course, in the presence of competing modes of diffusion the significance of such "velocities" is rather obscure. In general, the analysis of radiation-belt diffusion requires a complete application of the governing equations (see Chapter V). The direct identification of "diffusion velocities" from observations at fixed *energy* has enjoyed some historical popularity (see Section IV.6), but is no longer regarded as an adequate quantitative treatment of observational data.

IV. Prototype Observations

IV.1 Preliminary Considerations

Measurements of radiation-belt particle fluxes often reveal temporal variations that are unaccompanied by comparable variations of the local geomagnetic field. These temporal variations definitely indicate violation of one or more of the adiabatic invariants. For example, following large enhancements in intensity above the mean level characteristic of a specific particle energy, drift shell, and pitch angle, particle fluxes often exhibit an approximately exponential decay, apparently caused by *pitch-angle diffusion* into the loss cone. (Pitch-angle diffusion violates the first and/or second adiabatic invariant.) Large depletions of flux (resulting from nonadiabatic processes that operate during substorms) often are erased in time by a diffusive exchange of trapped particles among drift shells. This latter process, known as *radial diffusion*, can be identified with violation of the third invariant.

It must be presumed that pitch-angle scattering and cross-L diffusion do not terminate after attainment of a steady state in the radiation-belt flux distribution. Rather, these invariant-violating processes undoubtedly persist (perhaps with modified intensity) as permanent phenomena of radiation-belt dynamics. Thus, observation of the steady-state flux distribution reveals spatial structures that can be considered to arise from a dynamical balance among competing nonadiabatic source and loss processes.

The observations considered in the present chapter include examples from both of the general categories outlined above, *i.e.*, temporally varying and temporally static. Some of the particle observations reveal drift-phase organization (a consequence of processes that violate only the third adiabatic invariant). Drift-phase organization is manifested in the common occurrence of "drift-periodic echoes" in energetic proton and electron fluxes. Such events demonstrate that radial diffusion must occur. Examples of the electromagnetic and electrostatic disturbances apparently responsible for the various dynamical processes have also yielded to direct observation in recent years. These observations, when made in coincidence with measured temporal variations of the particle fluxes, indicate (for example) that particle-precipitation events are often accompanied by bursts of electromagnetic noise. Thus, the representative observations compiled in this chapter illustrate the observational evidence upon which rests the current understanding of radiation-belt dynamics.

The observations summarized in this chapter and elsewhere in the volume have involved a variety of spacecraft. For convenience, the orbital characteristics of these spacecraft are listed in Table 9 [main source: *TRW Space Log* **10**, 84 (1972)].

Table 9. Orbital Data for Cited Spacecraft

Name of Satellite	Date of Launch	Min r/a	Max r/a	Inclination	Period, minutes	Figure Numbers
Alouette 1	28 Sept 62	1.16	1.16	80.5°	105.4	51, 52
Alouette 2	28 Nov 65	1.08	1.47	79.8°	121.4	51
ATS 1	6 Dec 66	6.59	6.59	0.2°	1436.	b
Elektron 3	11 July 64	1.06	2.10	60.9°	168.	55
ERS 13	17 July 64	1.03	16.39	36.7°	2352.	1
Explorer 7	13 Oct 59	1.09	1.17	50.3°	101.2	51
Explorer 12	15 Aug 61	1.05	13.12	33.3°	1585.	46, 47
Explorer 14	2 Oct 62	1.04	15.46	32.9°	2184.	1, 22, 52, 79
Explorer 15	27 Oct 62	1.05	3.72	18.0°	312.0	c
Explorer 26	21 Dec 64	1.05	5.11	20.2°	456.	9, 38—41, 43
Explorer 34	24 May 67	1.04	34.14	67.1°	6231.	42
Gemini 4	3 June 65	1.03	1.04	32.0°	89.0	71
Gemini 7	4 Dec 65	1.03	1.05	28.9°	89.2	71
Hitch-Hiker 1	26 June 63	1.05	1.65	82.1°	132.6	49
Injun 1	29 June 61	1.13	1.16	67.0°	103.8	32
Injun 3	12 Dec 62	1.04	1.44	70.3°	116.3	1, 51, 57, 59
Injun 4[a]	21 Nov 64	1.09	1.39	81.4°	116.3	48
Injun 5[a]	8 Aug 68	1.11	1.40	80.6°	118.3	48
OGO 1	4 Sept 64	1.04	24.45	31.1°	3840.	1
OGO 3	6 June 66	1.04	20.14	30.9°	2907.9	50, 56
OGO 4	28 July 67	1.06	1.14	86.0°	98.1	48
OV1-2	5 Oct 65	1.06	1.54	144.3°	127.5	73
OV1-14	6 Apr 68	1.09	2.56	100.0°	207.8	48
OV3-3	4 Aug 66	1.06	1.70	81.6°	136.6	48
OV3-4	10 June 66	1.10	1.74	40.8°	143.2	82
Starad	26 Oct 62	1.03	1.87	71.4°	147.8	1
Telstar 1	10 July 62	1.15	1.88	44.8°	157.8	33, 41
1962-AΥ1	1 Sept 62	1.05	1.11	82.8°	94.4	71
1963-38C	28 Sept 63	1.17	1.18	89.9°	107.4	1, 38, 40
1963-42A	29 Oct 63	1.04	1.06	89.9°	90.9	1, 35, 36, 71
1964-45A	14 Aug 64	1.02	1.05	95.5°	89.0	1, 51, 71

[a] Injun 4 ≡ Explorer 25; Injun 5 ≡ Explorer 40
[b] ATS-1 data: Figs. 18, 23, 41, 42, 44, 45, 50, 60, 63, 64, 66, 67
[c] Explorer-15 data: Figs. 34, 41, 53, 54, 58, 73—78, 80, 81

IV.2 Decay of Particle Flux (Inner Zone)

Electrons. Electrons injected into the inner radiation zone by a high-altitude nuclear explosion (Starfish) on 9 July 1962 have provided an unusual opportunity to measure and study the natural decay of particle fluxes as a function of time. Unfortunately, the satellite best instrumented to measure these inner-zone fluxes (Telstar 1, a communications satellite) was launched on the day following detonation, and so only meager information on the natural (pre-Starfish) inner-zone electron radiation is available. Starfish electrons have only recently decayed sufficiently to permit measurements of the natural inner belt and its temporal variations (see Section IV.6).

Observations of Starfish electrons have been made at low altitudes and low L values by instruments on the satellites Injun 1 and Injun 3. The relevant measuring instrument on each satellite was a small, heavily shielded Geiger tube which detected the electrons from their intermediate bremsstrahlung in the counter shield. Temporal observations of the post-Starfish omnidirectional electron fluxes ($E \gtrsim 2\,\text{MeV}$) at $L < 1.30$ are plotted in Fig. 32 for several values of the total field intensity B (corresponding to several different altitudes above the earth) [74]. The plotted data points are not raw counting rates, but have been corrected for the Starfish-produced enhancement of the proton background (see below). These corrections proved to be significant only for $B > 0.20$ gauss.

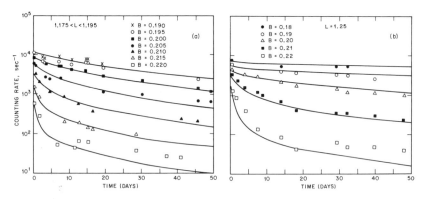

Fig. 32. Decaying inner-zone electron fluxes ($E \gtrsim 2\,\text{MeV}$) observed on Injun 1 following Starfish high-altitude nuclear explosion (9 July 1962). Solid curves are predictions based on atmospheric-scattering theory for $L = 1.185$ and $L = 1.25$, respectively [74].

The solid curves in Fig. 32 were determined theoretically by assuming that the artificially injected electron population experienced decay solely in consequence of atmospheric collisions (see Section II.2). Although some uncertainties may exist in deducing the proton subtractions that ultimately yield the data points in Fig. 32, the success in fitting the observed decays with the atmospheric-scattering model suggests that other redistributive processes were comparatively insignificant during the fifty-day interval illustrated.

Instruments on the Telstar satellite made measurements of the near-equatorial inner-zone electron fluxes beginning the day after the Starfish injection (see above). The measurements were made with silicon p-n junction solid-state detectors mounted behind various entrance-collimator geometries and shield thicknesses. Data for several L values ($E > 0.5$ MeV) are shown in Fig. 33 beginning on the day following

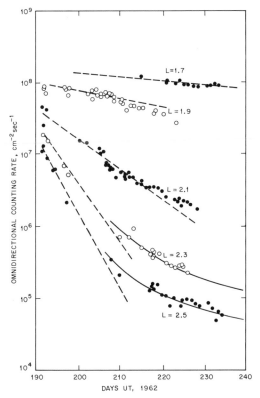

Fig. 33. Decaying inner-zone electron fluxes ($E \gtrsim 0.5$ MeV) observed near equator on Telstar 1 following Starfish event (Day 190). Solid curves and dashed lines are empirical [75].

the Starfish explosion [75]. The data were taken close to the equator at $L = 1.7$ and $L = 1.9$. Because of the orbital inclination, measurements at the higher L values were made at progressively higher latitudes (*e. g.*, latitude 30° at $L = 2.5$).

After the explosion (Day 190) the electron fluxes exhibited approximately exponential decay until about Day 230. The apparent deviation from exponential decay at the higher L values ($L \geq 2.3$) and the later times (from approximately Day 210) actually arises from the response of the detector to inner-zone proton contamination. The lifetimes extracted from the initial exponential decays are much shorter at the higher L values than those predicted by an atmospheric-scattering model [42]. Indeed, the decay times actually *decrease* with increasing L, in explicit contrast to the prediction of any reasonable atmospheric-scattering model, *i. e.*, lifetimes *increasing* with altitude. These observations were among the first to suggest that wave-particle interactions are important in controlling magnetospheric particle dynamics.

The pitch-angle distribution of electrons injected by a high-altitude nuclear explosion is typically abnormal in the sense that few of the particles mirror near the equator (unless, of course, the detonation is equatorial). A very interesting observational result based on the USSR detonation of 28 October 1962 (Day 301) is the restoration of a pitch-angle distribution to the fundamental mode of pitch-angle diffusion. The electron measurements confirming this phenomenon were made with a shielded solid-state *p-n* junction detector flown on the near-equatorial satellite Explorer 15. Decay of the initially anomalous pitch-angle distribution to its lowest normal mode is illustrated sequentially in Fig. 34 [43]. Here the electron omnidirectional counting rates beginning with Day 301 are plotted as a function of $X \equiv [1 - (B_e/B)]^{1/2}$, where B_e is the equatorial magnetic-field intensity. Most electrons injected by the nuclear blast appeared initially with equatorial pitch-angle cosine $\gtrsim 0.6$; the omnidirectional electron flux at $L = 1.9$ initially exhibited an off-equatorial maximum (at $X \approx 0.8$). The initial distribution of equatorial pitch angles can be viewed as a superposition of normal modes. Each normal mode decays at a characteristic rate, with the longest lifetime belonging to the lowest eigenmode. This lowest mode is the normal pitch-angle distribution toward which the initial distribution decays. The off-equatorial peak in omnidirectional flux disappears rather quickly (in about ten days), and the lowest normal mode is essentially isolated after about thirty days. The entire pitch-angle distribution, operating in its lowest normal mode, then decays with a particle lifetime ~ 40 days.

The decay of a naturally occurring, nearly monoenergetic, electron enhancement near the inner edge of the inner belt was observed during

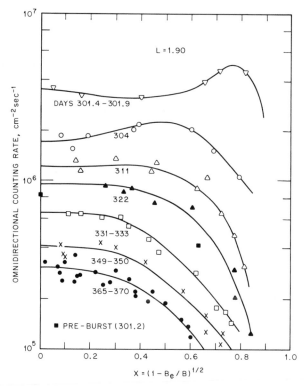

Fig. 34. Decaying omnidirectional intensities of inner-zone electrons ($E > 1.9$ MeV) observed on Explorer 15 following nuclear event of 28 October (Day 301) 1962. Solid curves are empirical [43].

a one-week period in 1963 following a magnetic storm on October 30 [76]. The relevant data were obtained with a plastic scintillation spectrometer flown on the earth-oriented, polar-orbiting satellite 1963-42A. The observed spectral peak was centered at approximately 1.3 MeV and exhibited a full width ≈ 0.23 MeV at half maximum (approximating the limit of instrumental resolution). The approximately exponential decay (with time) of the peak intensity at each of several L values is illustrated in Fig. 35 and compared there with the decay predicted by atmospheric-scattering theory. As was observed in the decay of the Starfish electrons (Fig. 32), the decay rates at these low L-values are consistent with the interpretation that electrons are lost solely through collisions with atmospheric constituents.

The apparent displacement (≈ 440 km) of the dipole axis from the earth's center permits the observation of particles locally trapped by

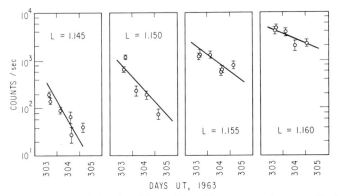

Fig. 35. Decaying omnidirectional intensities of inner-zone electrons ($E > 1.3$ MeV) observed on 1963-42A following a natural "injection" event on 30 October (Day 303) 1963. The range in B is 0.2119—0.2222 gauss. Solid curves are predictions based on atmospheric-scattering theory [76].

the magnetic field but doomed ultimately to precipitate. The 440-km displacement contributes to an "anomalously" weak-field condition (≈ 0.25 gauss, vs 0.31 gauss in the centered-dipole model) at sea level in the South Atlantic region. Many inner-zone drift shells thus plunge deep into the atmosphere in the vicinity of the "anomaly", while tracing a constant mirror-field intensity. Particles locally trapped on such drift shells are destined to precipitate as they drift into the "anomaly" region. Most of the actual inner-zone electron precipitation occurs here, although the pitch-angle diffusion responsible for this precipitation is distributed among all magnetic longitudes.

Indirect studies of the L-dependent decay rates of inner-zone electrons have been made by studying azimuthal variations in the low-altitude fluxes of precipitating electrons. The intensities of electrons that will mirror below the 100-km altitude exhibit significant variations with longitude. These azimuthal variations are attributed to the pitch-angle diffusion produced by atmospheric scattering and wave-particle interactions as the electrons drift from west to east. The fluxes are found to be higher to the west of the "anomaly" than to the east of it.

The general increases in electron intensity with increasing east longitude beyond the "anomaly" suggest an L-dependent pitch-angle diffusion process operating in the presence of a contracted loss cone. On each L shell, restoration of pitch-angle isotropy via diffusion is prevented only by precipitation into the atmosphere, which typically occurs at the "anomaly" (where the loss cone abruptly expands). Measurements of the electron flux at various drift phases of an adiabatic drift shell can thus be used to extract a pitch-angle diffusion coefficient. Examples of the azimuthal variation of the fluxes of electrons ($E > 0.4\,\text{MeV}$) that would mirror in the "anomaly" at altitudes $< 100\,\text{km}$ are shown in Fig. 36 [77]. These data, obtained with a plastic scintillation detector attached to a photomultiplier flown on the polar orbiting vehicle 1963-42A, are plotted as a function of longitude measured (in degrees) eastward from the South Atlantic "anomaly".

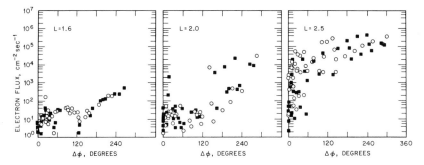

Fig. 36. Daytime (open circles) and nighttime (filled squares) measurements of omnidirectional flux of electrons ($E > 0.4\,\text{MeV}$) that would mirror in the "anomaly" ($\Delta\varphi = 0$) at altitudes $< 100\,\text{km}$ [77], obtained at various magnetic longitudes (φ) on three separate B,L contours.

Protons. High-energy protons trapped in the earth's magnetic field and mirroring at low altitudes were first detected (and their energy spectra measured) by the use of recoverable nuclear-emulsion packets flown on rockets and satellites. A large number of these emulsion packets were flown during the years 1961—64, yielding much data on the temporal decay of the Starfish-produced proton fluxes.

Recovered emulsion records indicating the flux of protons ($E \sim 55\,\text{MeV}$) in the South Atlantic "anomaly" were divided into altitude increments and then examined as a function of time. The proton intensities for altitudes of $440\,\text{km}$ and $350\,\text{km}$ are shown in Fig. 37 [78]. The dashed vertical lines indicate the day of the Starfish explosion. Subsequent to July 1962, the Starfish-enhanced proton fluxes decayed steadily with time toward their pre-Starfish levels. The solid curves drawn through the data points represent the theoretically expected fluxes (above the natural level) of Starfish protons, assuming that ionization of the atmosphere (see Section II.2) is the only loss mechanism for these protons. The drift-averaged values [41] of atomic-electron densities $\Sigma Z_i \bar{N}_i$ [cf. (2.04)], using the Harris-Priester model atmosphere

[79] below 1000 km, are noted for each observation altitude in Fig. 37. The agreement between the observations and the theoretical predictions is very good. This result suggests that the ionization theory, incorporating the model atmosphere, is sufficient to account for the decay of Starfish-produced proton fluxes observed at these low altitudes.

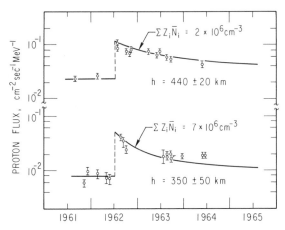

Fig. 37. Inner-zone omnidirectional proton flux ($E = 55$ MeV) measured at sporadic times in two altitude ranges above the South Atlantic "anomaly". Sharp increases (dashed lines) coincide with Starfish event (9 July 1962). Solid curves post-Starfish are predictions based on atmospheric-collision theory [78], assuming a source sufficient to maintain the mean pre-Starfish intensity (horizontal solid lines). Post-Starfish curves are normalized to the first data point measured at each altitude following the nuclear event.

IV.3 Decay of Particle Flux (Outer Zone)

Electrons. Extensive studies of the outer radiation belt have shown that decay begins to predominate two or three days after a magnetic storm produces enhancements of the electron flux. The electron intensities then decay in an approximately exponential manner, as in the inner zone.

The typical temporal behavior of outer-zone electron fluxes (rapid increases during storms, followed by steady decays) can be seen readily in Fig. 38 where a four-month time history of electron fluxes ($E > 1$ MeV) is plotted at each of three different L values [80]. Both high-altitude (Explorer-26) and low-altitude (1963-38C) satellite data are plotted at each L value in order to compare the fluxes measured at different

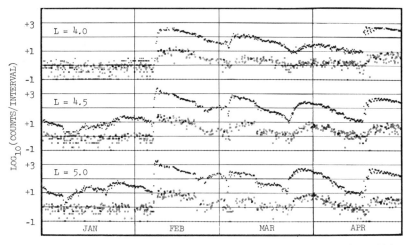

Fig. 38. Unidirectional counting rates of electrons observed on polar-orbiting satellite 1963-38C ($E > 1.2$ MeV, open squares) and on near-equatorial satellite Explorer 26 ($E > 1.0$ MeV, crosses) during four months in 1965 [80]. One count per interval represents a flux ≈ 1200 cm^{-2} sec^{-1} ster^{-1}.

locations in a given flux tube. The data were obtained with solid-state detectors.

The steady decay of flux observed after storm-produced enhancements is occasionally interrupted by short-term variations (either increases or decreases) that last for several hours to a day. These variations, superimposed on the long-term decay, have been identified as adiabatic modulations caused by gradual time variations of the local magnetic field intensity on a given L shell. These gradual changes in the local field intensity result primarily from temporal changes in the low-energy (~ 25-keV) proton density in the quiet-time ring current, whose density generally peaks at $L \sim 6$. Additional adiabatic modulation may arise from changes in the intensity of currents flowing on the magnetosphere boundary.

As in the case of inner-zone electron fluxes, the e-folding times extracted from the approximately exponential decays of outer-zone fluxes following the storm-time enhancements can be taken as estimates of electron lifetimes. The lifetimes measured for selected L values and energies following the storm of 18 April 1965 are shown in Fig. 39 [80]. Electron lifetimes are thus found to vary directly with energy and inversely with L in the outer radiation zone.

Natural enhancements of the electron flux, especially in the region $L \approx 3.0$—3.5, were observed in association with the large storm of 18

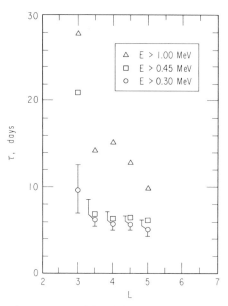

Fig. 39. Lifetimes of near-equatorial unidirectional electron fluxes, as determined from Explorer-26 data (see Fig. 38) for the period 22.5 April 1965 to 3.0 May 1965 [80].

April 1965. At both high and low altitudes in this region, the enhanced fluxes were observed to decay more quickly during the first few days following "injection" than during the subsequent quiet period (see Fig. 40). No similar temporal change in the decay rate is apparent outside this narrow interval of L [80]. In view of the nonlinear processes (Section II.6) that tend to act upon enhanced particle fluxes, it is perhaps quite reasonable to expect such behavior in the region where storm-associated particle injection is most intense.

The e-folding decay times of Fig. 39, determined from the flux decays following the storm of 18 April 1965, are comparable to the generally accepted lifetimes that enter most discussions of pitch-angle diffusion. This can be seen from the electron lifetimes plotted in Fig. 41. These decay times were measured from data obtained at various epochs by instruments on four different elliptically orbiting satellites [43] and on the synchronous satellite ATS 1 ($r = 6.6a$). The data were obtained over approximately a five-year period. The overall similarity of the lifetimes from one determination to another suggests that the basic magnetospheric processes acting to produce the pitch-angle diffusion persist from year to year.

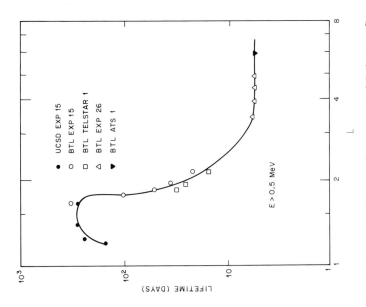

Fig. 41. Observed lifetimes of near-equatorial electron fluxes in the inner and outer zones [43].

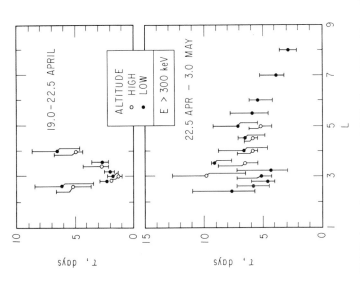

Fig. 40. Lifetimes of near-equatorial (Explorer 26) and low-altitude (1963-38C) unidirectional electron fluxes [80]. These lifetimes are determined for two successive time intervals following the magnetic storm of 18 April 1965 (see Fig. 38).

Protons. Pitch angle diffusion is not ordinarily known to affect radiation-belt protons at energies $E \gtrsim 1$ MeV. There exists no observational evidence to suggest that such protons fail to obey (3.57) deep in the inner magnetosphere ($L \lesssim 1.7$). In the outer magnetosphere, however, the fluxes of energetic protons ($E \sim 5$—70 MeV) are observed to rise and decay on time scales ~ 15 minutes. Such observations (see below) must be interpreted with caution in the context of particle diffusion, since solar-flare protons often apparently have free access to the synchronous orbit ($r = 6.6a$) where the observations [81] have been made.

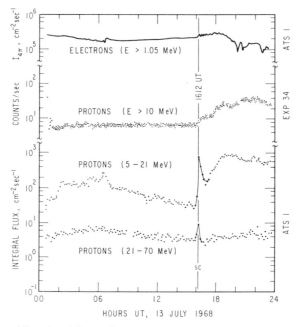

Fig. 42. Omnidirectional fluxes of solar protons and outer-belt electrons observed at synchronous altitude on ATS 1, and of solar protons ($E > 10$ MeV) observed simultaneously in interplanetary space on Explorer 34 [81].

In Fig. 42, a proton enhancement at synchronous altitude accompanies the sudden commencement (1612 UT, 13 July 1968) of a magnetic storm, and subsequently decays on a time scale ~ 15 min in the absence of further magnetic-field variation at synchronous altitude [81]. In a similar event on 20 November 1968 (see Fig. 67, Section IV.8), the decay was modulated by drift-periodic echoes (see Section III.1) in the proton flux, indicating the presence of closed drift orbits. In this observation, the proton population observed at the satellite near local

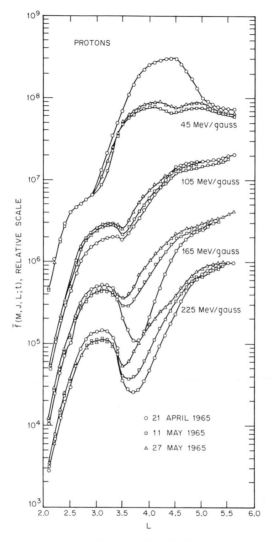

Fig. 43. Temporal evolution of proton distribution function at $J=0$ following magnetic storm of 18 April 1965 [82], based on Explorer-26 data.

midnight apparently does not have free egress to interplanetary space. The value of ε (see Section I.1) is rather large (~ 0.1) for the behavior of these protons to be interpreted in terms of adiabatic theory, however. Thus, there may be a fractional probability on each drift period for a proton to escape the magnetosphere.

The proton decay rate in Fig. 42 clearly increases with proton energy, as an interpretation based on the breakdown of adiabatic theory ($\varepsilon \gtrsim 0.1$) would lead one to expect. On the other hand, ground-based observations of storm-associated geomagnetic micropulsations suggest the presence of adequate spectral intensity near $\omega/2\pi \sim 1$ Hz to account for the rapid decay rates observed. In the magnetosphere, these Pc-1 micropulsations (ion-cyclotron mode) could interact with the protons through a Doppler-shifted cyclotron resonance (see Section II.5), thereby causing pitch-angle diffusion.

At proton energies $E \lesssim 0.5$ MeV, the evidence for pitch-angle diffusion is unequivocal. In Fig. 43, the interstorm (18 April to 15 June 1965) evolution of $\bar{f}(m,J,L;t)$ at $J=0$ is illustrated for selected values of M [82]. The pronounced decay between 21 April and 11 May at $M=45$ MeV/gauss illustrates the operation of a nonlinear mechanism (*cf.* Section II.6) for self-limiting of the trapped proton flux. Once the stable limit is achieved, the decay ceases (*cf.* 11—27 May). The contrasting behavior of $\bar{f}$ at $M=225$ MeV/gauss and $L \gtrsim 3.5$ is evidence for radial diffusion (inward from a source beyond $L \approx 5$ and outward from an apparent source at $L \approx 3$). The distribution function clearly increases with time in the region $3.5 \lesssim L \lesssim 4.5$ so as to fill the "slot" between regions of higher radiation intensity (*cf.* Section IV.6).

IV.4 Statistical Observations

Extended observations of the electron-radiation environment at synchronous altitude have been carried out from the satellite ATS 1. A compilation of ATS-1 data on the integral omnidirectional flux $I_{4\pi}$ above four energy thresholds is illustrated in Fig. 44 [83]. The results are presented in the form of a probability P that the flux $I_{4\pi}$ will exceed a given level Q. For $0.1 \lesssim P \lesssim 0.6$, this probability is quite linear with respect to $\log Q$. Extrapolation of the linear fits to $P=1$ and $P=0$ permits definition of a "maximum" flux I^* and a "minimum" flux I_{min} for each energy threshold.

Two properties of Fig. 44 are immediately apparent. First, the value of $\log I^*$ increases with decreasing energy threshold. Second, the range of probable fluxes, as measured by $\Delta \log I \equiv \log(I^*/I_{min})$, narrows with decreasing energy. These properties are shown quantitatively in Fig. 45. Extrapolation of the results to $E=40$ keV [*cf.* (2.69a) for $s=0.5$, $B=125\gamma$, $N_e=1$ cm^{-3}] yields $I^*(E^*) \sim 10^8$ cm^{-2} sec^{-1}, in good agreement with (2.72). Moreover, the value of $\Delta \log I$ is approximately 0.18 at $E=E^*=40$ keV, *i. e.*, the probable fluxes (as identified by linear extrapolation in Fig. 44) lie within the range $0.6I^*$ to $1.0I^*$ at $E=E^*$.

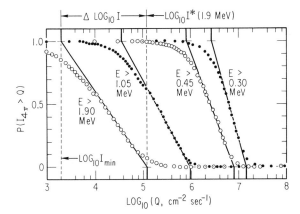

Fig. 44. Compilation of ATS-1 electron data [83], showing the probability P that the integral omnidirectional flux $I_{4\pi}$ exceeds a specified value Q. Solid lines empirically fit the data points for $0.1 \lesssim P \lesssim 0.6$.

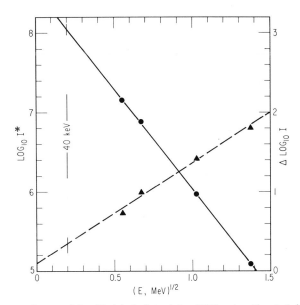

Fig. 45. Dependence of $\log I^*$ (circles) and $\log(I^*/I_{min}) \equiv \Delta \log I$ (triangles) on threshold energy E. Solid lines empirically fit the intercepts extracted from Fig. 44. The unusual choice of abscissa facilitates extrapolation to the lower energies.

The self-limiting property of synchronous-altitude electron fluxes thus enforces a very narrow range of flux variability at $E \approx E^*$. The magnetosphere apparently becomes increasingly tolerant of integral-flux variability for particle energies well above E^*, the threshold for cyclotron resonance with a growing wave.

IV.5 Static Flux Profiles

In the absence of temporal variations in the observed particle fluxes, it is frequently possible to obtain information about competing dynamical processes from observations of the steady-state distributions. In view of the multiplicity of unobserved processes that may be competing, such steady-state distributions must always be interpreted with special caution. However, there are a number of steady-state observations that suggest a prominent role for radial diffusion at constant M and J.

Protons. Extensive measurements of the lower-energy proton distributions (100—500 keV) in the magnetosphere have been made by scintillation-counter detectors flown on the satellites Explorer 12, 14, 15, and 26. These particle populations are found to be very stable in time and to have energy spectra that are exponential over a wide range of energy and L. Further, the spectra harden with decreasing L. Three spectra obtained at different L values in 1961 on Explorer 12 are shown

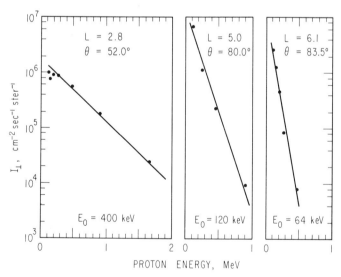

Fig. 46. Outer-zone proton spectra observed [84] on Explorer 12 (26 August 1961).

in Fig. 46 [84]. More recent magnetospheric proton spectra for energies between 0.17 MeV and 3.4 MeV have been obtained from an instrument on satellite 1964-45A in 1964 and 1965 [85]. Where comparisons are possible, these data agree with those obtained earlier (in 1961).

It is found empirically from data such as those in Fig. 46 that E_0, the e-folding energy of the exponential proton spectrum, varies as L^{-3}; *i. e.*, in direct proportion to the equatorial field intensity B_e. This variation of E_0 with B_e is often cited as evidence for radial diffusion at constant M and J. Such a claim probably overstates the case. What can be shown is that any steady-state ($\partial \bar{f}/\partial t = 0$) solution of (3.01) (the equation governing radial diffusion at constant M and J, excluding other dynamical processes) has the property that $(\partial \ln \bar{f}/\partial \ln p)_{L,y}$ is independent of L for particles having the same J^2/M if D_{LL} and all boundary locations are independent of particle energy [33].

For nonrelativistic particles ($E = p^2/2m_0$) it is convenient to represent a power-law energy spectrum as $J_\perp(E) = J_\perp(p_0^2/2m_0)(p_0/p)^{2l}$ and an exponential spectrum as $J_\perp(E) = J_\perp(0)\exp(-p^2/2m_0 E_0)$. The statement concerning the L-independence of $(\partial \ln \bar{f}/\partial \ln p)_{L,y}$ in the steady state implies that

$$(\partial \ln \bar{J}_\perp/\partial \ln p)_{L,y} = 2 + (\partial \ln \bar{f}/\partial \ln p)_{L,y} = \text{constant} \qquad (4.01)$$

on a surface of constant M and J in phase space. The power-law spectrum retains its form with p_0 and l independent of L. The exponential spectrum retains its exponential form, but E_0 varies with L in such a manner that $E_0(L) \propto 1/y^2 L^3$. In other words, the e-folding energy of the exponential spectrum behaves like the energy of an individual particle.

In view of the energy dependence of competing dynamical processes (including radial diffusion caused by electrostatic impulses; *cf.* Section III.3) it is quite remarkable that the observations of E_0 satisfy (4.01), even at $y = 1$. Even more remarkable is that the observations appear to satisfy (4.01) at all other values of y. In Fig. 47 the observed e-folding energy E_0 is plotted against L for selected values of $\sin^{-1} y_7$ [86]. The "theoretical" curves appear to converge on a common value of E_0 at $L \approx 10$. This latter location can perhaps be interpreted as the location of the proton source.

These data at most *suggest* the inward radial diffusion of the protons to lower L values from an external source. In view of possibly competing energy-dependent processes, the data could perhaps be interpreted in other ways. Furthermore, the application of (4.01) yields no specific

---

[33] Under these stringent conditions, the solutions $\bar{f}$ corresponding to two distinct values of E_7 (see Fig. 24, Section III.1) can differ only by a multiplicative factor independent of L, in view of the "uniqueness" of steady-state solutions to (3.01).

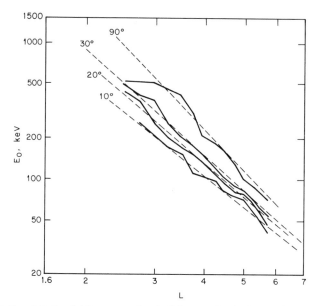

Fig. 47. Empirical e-folding energies (solid curves) of observed proton spectra [84] at y values consistent with (3.02); expected variation (dashed curves) of energy with L for individual protons having constant M and J, at selected values of E_7 and $\sin^{-1} y_7$ [86].

magnitude for the diffusion coefficient allegedly responsible for establishing the observed distributions. The radial transport coefficient must be obtained from considerations other than the time-independent particle data.

Heavier Ions. The presence of alpha particles (and heavier ions), as well as protons, in the radiation belts offers a possible opportunity to select a dominant radial-diffusion mechanism from Chapter III. Conventionally, fluxes of alpha particles and protons are compared at the same energy per nucleon (see Section III.3) so as to obtain the alpha-to-proton (α/p) ratio as a function of L. Values of this ratio are indicated [34] for $L \sim 3$ in Fig. 48 [59]. An energy of ~ 0.5 MeV at $L \sim 3$ scales (at constant M and J, with $J=0$) to ~ 10 keV at $L \sim 10$, where interplanetary ions may be expected to enter the magnetosphere.

At the same energy/nucleon, the solar-wind [9] and solar cosmic-ray [88] α/p ratios (~ 1—20×10^{-2}) are substantially larger than the values

---

[34]The decay of the α/p ratio (Fig. 48) between 8 November 1968 and 26 January 1969 follows an enhancement accompanying the magnetic storm of 1 November 1968.

~ 1—20 × 10^{-4} plotted in Fig. 48[35]. If the magnetospheric helium-ion source is solar[36], then helium nuclei are either lost more quickly than protons at a given L value or they diffuse inward more slowly. A consideration of possible loss mechanisms (cf. Sections II.2 and II.5) tends to rule out the former possibility. Since a helium ion with the same energy/nucleon as a proton has a larger azimuthal-drift speed than the proton, helium ions would tend to diffuse inward (under electrostatic fluctuations with an ω^{-2} spectral density) more slowly than pro-

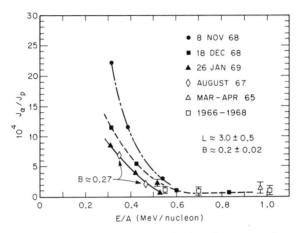

Fig. 48. Low-altitude observations of α/p ratio (based on comparison of $J_\perp$ values) at several different epochs [87].

tons. There would be no difference among radial diffusion coefficients caused by an inverse-square *magnetic* power spectrum. Radial diffusion of solar helium ions by electrostatic (rather than electromagnetic) impulses (see Section III.3) thus apparently can account for the reduced helium-to-proton ratio observed in the inner magnetosphere [40]. This latter conclusion depends upon assumptions concerning the ion source and cannot be extrapolated without thought to all energies. However, any indication of radial diffusion driven by electrostatic impulses tends to cast doubt on the customary interpretation of Figs. 46 and 47.

_____

[35]The values shown in Fig. 48 were obtained from low-altitude (high-latitude) observations. Recent equatorial measurements [87] suggest an α/p ratio ~ 1—20 × 10^{-2} at E/A ~0.4 MeV/nucleon, for $3.1 \lesssim L \lesssim 4.4$.

[36]Alternatively, the source for magnetospheric helium radiation might be ionospheric, in the form of He$^+$ ions blown into the earth's plasma sheet by the polar wind [89].

Electrons. It is much more difficult to study the spectral and spatial parameters of outer-zone electron fluxes than of proton fluxes, since the former vary so drastically with time (*cf.* Fig. 38). However, several studies of this nature are on record. In one such study the spectral parameter E_0 for electrons ($E = 1—4$ MeV) was obtained from data measured by a scintillation spectrometer on the low-altitude satellite Hitch-Hiker 1 during July and August 1963. The results are plotted in Fig. 49 as a function of L for several days in July [90]. The solid line indicates an empirical hardening of the spectrum with decreasing L, given by $L^{1.3} E_0 = 2.4$ MeV.

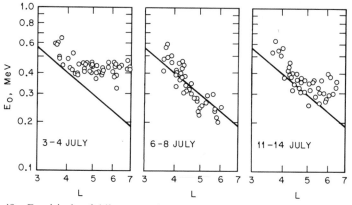

Fig. 49. Empirical *e*-folding energies (open circles) of electron spectra ($E = 1—4$ MeV) observed during three periods in July 1963 [90]. Coverage on 4 July extends from 00 UT to 03 UT; the other eight days have 24-hour coverage. Solid lines correspond to $L^{1.3} E_0 = 2.4$ MeV.

During the period of 3—4 July, the spectral parameter evidently became L-independent beyond $L \sim 4$. During the geomagnetically more active period of 6—8 July, the spatial variations in the spectral parameter E_0 agreed qualitatively with that "expected" from (4.01). The distinction between quiet and disturbed times could possibly be taken to mean that radial diffusion at constant energy (via drift-shell splitting; see Section III.7) dominates that at constant M and J during quiet intervals.

The phenomenon of drift-shell splitting has been verified by a comparison of data obtained with electron spectrometers on the ATS-1 and OGO-3 satellites [91]. The diurnal variation of electron fluxes (higher at noon than at midnight) and pitch-angle distribution ($J_{\pi/2} > J_{\pi/3}$ at noon; $J_{\pi/3} > J_{\pi/2}$ at midnight) observed at fixed energy on ATS 1 are found to be adiabatically compatible (at least during quiet periods)

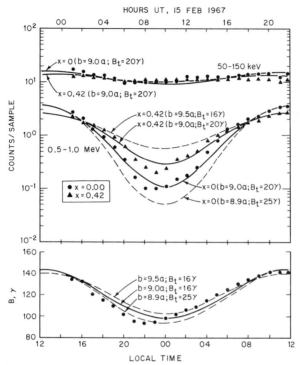

Fig. 50. Diurnal variation of electron fluxes and magnetic field (data points) observed at synchronous orbit (ATS 1), as compared with predictions (dashed and solid curves) based on Mead-Williams field models $(b/a, B_t)$ and distribution function $\bar{f}(M, J, \Phi)$ deduced from OGO-3 electron data [91].

with the flux profiles obtained over a broad range of L values from the OGO-3 data (see Fig. 50). Although drift-shell splitting is thus established as an adiabatic phenomenon (see Section I.7), its quantitative influence on radial diffusion can be determined only by inserting a pitch-angle diffusion coefficient (either measured directly, or deduced from an exponential decay rate) in (3.37). When this is done (see Section IV.6), it is found that shell splitting effects can seldom account for even 10 % of the observed radial diffusion. The correct interpretation of Fig. 49 is probably other than one based on constant-energy diffusion.

Finally, Fig. 51 illustrates the yearly "history" of the outer electron belt during a period of decreasing solar and geomagnetic activity (decreasing K_p) in terms of the "crest" position L_c of three integral unidirectional-flux profiles [92]. These annual-mean crest positions could be regarded as quasi-static features of the radiation belt, resulting from the balance between loss (pitch-angle diffusion) and transport (radial diffusion) pro-

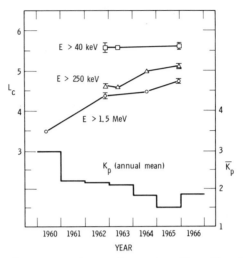

Fig. 51. Variation of outer-zone electron-flux crest positions ($L_c \equiv$ shell correspond-ing to maximum electron $I_\perp$) during a period of decreasing geomagnetic activity [92], as indicated by $\bar{K}_p$ ($\equiv$ annual mean of K_p).

cesses operative during the year. The essential difficulty in interpreting Fig. 51 is that the flux profiles are considered for fixed energy, whereas the radial-diffusion process is believed to operate at constant M and J. It is preferable, therefore, to examine profiles of $\bar{f}(M,J,L)$ at fixed M and J (*cf.* Fig. 43, Section IV.3).

IV.6 Time-Varying Flux Profiles

The most convincing evidence for the radial diffusion of radiation-belt particle fluxes has arisen from studies of temporal variations in the flux profiles (distributions in L). Most frequently, these studies examine the changes in the trapped-electron fluxes following storm-time enhance-ments. Most of the reported inward movements of flux profiles observed in the trapped-electron intensities have been especially evident in the higher-energy ($E \gtrsim 1$ MeV) fluxes. This last observation may be under-stood qualitatively by recalling the observed energy dependence of elec-tron lifetimes plotted in Fig. 39. Although radial diffusion could well be important for electrons of energy $E \sim 0.5$ MeV, their observed temporal development may be dominated by the persistent decay due to pitch-angle diffusion. The higher-energy electrons, with longer lifetimes, then survive to exhibit the strongest evidence for radial diffusion.

The first, and probably most widely quoted, observation of the inward movement of an energetic-electron flux profile was that reported for the period following the magnetic storm of 17—18 December 1962. The flux measurements of near-equatorial electrons ($E \gtrsim 1.6$ MeV) were made with a Type-302 Anton Geiger counter on the satellite Explorer 14, whose orbit is inclined by $33°$ to the earth's equatorial plane.

The flux profiles observed on several days following the storm are shown in Fig. 52a [111]. The inward radial movement of the "leading edge" of the electron distribution is clearly evident. An inward "velocity" $dL/dt \sim -0.04$/day at $L = 3.7$ is apparent in the data.

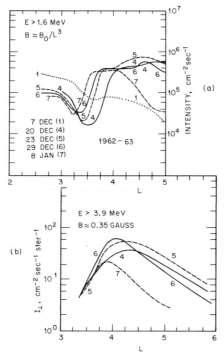

Fig. 52. Radial profiles of electron $I_{4\pi}$ [111] for $E > 1.6$ MeV near equator (a) and $I_{\perp}$ [112] for $E > 3.9$ MeV at low altitude (b), as observed on several days preceding and following the magnetic storm of 17 December 1962 (data from Explorer 14 and Alouette 1, respectively).

Plotted in Fig. 52b are the flux profiles of electrons ($E > 3.9$ MeV) detected during the same period from a shielded Type-302 Anton Geiger counter flown on the low-altitude satellite Alouette 1 [112]. The inward movement of the flux profiles measured on the low-altitude satellite

(Fig. 52b) is not as pronounced as that observed for particles mirroring closer to the equator. This observation perhaps confirms the theoretical prediction (see Section III.2) that radial diffusion caused by magnetic impulses will act primarily on particles that mirror near the equator.

A diffusion process known to conserve M and J, however, is not easily visualized in terms of constant-energy data as in Fig. 52. As indicated in the previous section, the phase-space distribution function $\bar{f}(M,J,L;t)$ should be plotted against L at selected intervals of time, with M and J held constant. Such a representation, however, requires information concerning the electron energy spectrum. Thus, data must be available from at least two (and preferably several) energy channels in the energy interval of interest.

The satellite Explorer 15 carried a pair of solid-state detectors able to measure the local unidirectional fluxes for $E > 0.5$ MeV and $E > 1.9$ MeV during the December 1962 event. Daily-median profiles of the $J = 0$ fluxes are plotted in Fig. 53 [93] for the same days as in Fig. 52. It is quite evident from these Explorer-15 data that the lower-energy electron flux shows temporal variations dominated by a steady decay of intensity following the storm. In contrast, the higher-energy flux exhibits the inward movement of a "leading edge" of the flux profile apparent in Fig. 52a. It is clear, then, that an adequate analysis of these data entails a thorough consideration of both of these superficially disparate observations. The first step in such an analysis is to present the data in the form of $\bar{f}(L,t)$ at constant M and J.

For equatorially mirroring ($J = 0$) particles, as represented in Fig. 53, the transformation from flux to $\bar{f}(L,t)$ is simplified by the fact that $\bar{f} = (1/2m_0 M B_0) L^3 \bar{J}_\perp$. The energy corresponding to each value of M is determined as a function of L from (2.07). The *differential* flux $J_\perp$ at each L-dependent energy is deduced by fitting a suitable spectral form (typically exponential or power-law) between the two measured (daily median) *integral*-flux values $I_\perp$ (*cf.* Fig. 53). Results of this interpolation are illustrated in Fig. 54 [93]. No "crest" is evident in the distribution function $\bar{f}(L,t)$ between $L = 3.4$ and $L = 4.8$; the particles in Fig. 54 appear to be diffusing inward from a source beyond $L \sim 5$. The superimposed decay (*cf.* Fig. 39, Section IV.3) appears to be greater at $M = 300$ MeV/gauss than at $M = 750$ MeV/gauss, and so the inward diffusion of $\bar{f}(L,t)$ at the lower M value is obscured. Quantitative procedures for extracting D_{LL} and τ from the data in this format are described in Chapter V.

A number of subsequent measurements showing an apparent inward movement of the "leading edge" of an electron-flux profile have been reported by various investigators. The data shown in Fig. 55 were obtained on the satellite Elektron 3 in September 1964 [94]. The fluxes

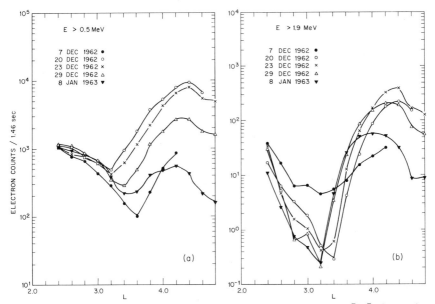

Fig. 53. Daily-median intensities of equatorially mirroring electrons [93] observed on Explorer 15 before and after the magnetic storm of 17 December 1962 (*cf.* Fig. 52).

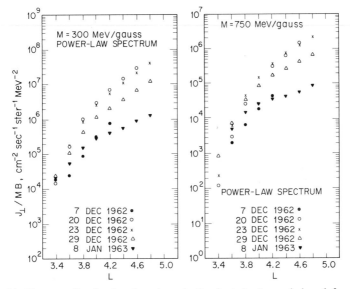

Fig. 54. Electron distribution functions ($\times 2m_0$) at $J=0$, as deduced from Fig. 53 and calibration data by assuming a power-law spectrum for energy interpolation [93].

labeled VS-1 are electrons of energy $E > 100\,keV$ and were measured with a shielded gas-discharge counter; the fluxes labeled VF-1 are electrons of energy $E > 400\,keV$ and were measured with a shielded NaI(Tl) crystal attached to a phototube. Inward movement of the "leading edge" in Fig. 55 is much less evident at $E \sim 100\,keV$ than at $E \sim 400\,keV$. This comparison is not surprising in view of Fig. 53, where inward movement is not apparent even at $E \sim 500\,keV$. Conditions in the radiation belts must have been different in September 1964 from those in December 1962 in order to permit detection of the apparent inward movement at $E \sim 400\,keV$. The 400-keV channel in Fig. 55 exhibits an apparent "leading-edge velocity" $dL/dt \sim -0.2/\text{day}$ at $L \approx 4.4$; as in the December 1962 event, a quantitative interpretation of the underlying dynamics would require the data to be converted to a distribution function at constant M and J.

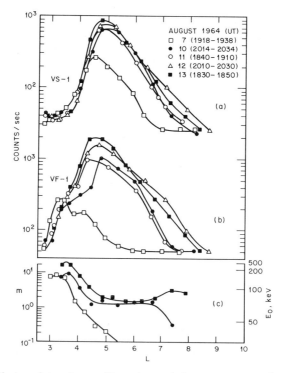

Fig. 55. Electron-intensity profiles observed in two energy channels (VS-1, $E > 100\,keV$; VF-1, $E > 400\,keV$) on Elektron 3 during August 1964, with spectral indices $m \equiv (-d\ln J_{4\pi}/d\ln E)$ and $E_0 = (-d\ln J_{4\pi}/dE)^{-1}$ deduced from ratio of counting rates [94].

Storm-associated enhancements of inner-zone electron fluxes became apparent beginning in mid-1966, by which time the artificial belts created by high-altitude nuclear detonations no longer significantly contaminated the inner zone (see Section IV.2). Electrons naturally "injected" at $L \sim 2$ subsequently appear to diffuse deeper into the inner zone. An example of the apparent inward movement of naturally "injected" electrons following the magnetic storm on 2 September 1966 is shown in Fig. 56 [95]. These data were obtained by a magnetic spectrometer

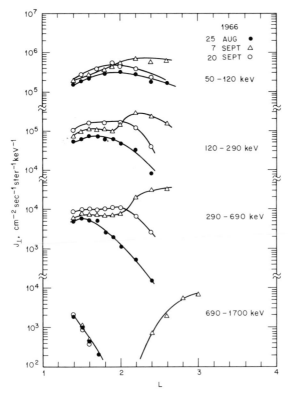

Fig. 56. Inner-zone electron-flux profiles observed on OGO 3 before and after the large magnetic storm of 2 September 1966 [95].

flown on the OGO-3 satellite. Electron fluxes were enhanced at all observed energies and L values in consequence of the September 1966 storm. The largest enhancements appeared in the interval between $L \approx 2.0$ and $L \approx 2.4$. On the eighteenth day after the storm the electron intensities at the lower L values were continuing to increase, while the original enhancements at $L \approx 2.0$—2.4 had disappeared.

It is essentially impossible to identify a "leading-edge velocity" in Fig. 56, but a "crest velocity" $d L_c/dt \approx -0.04$/day can perhaps be deduced at 50—120 keV. This "velocity" should be assigned to the median L_c value ($L \approx 2.1$) at which $(\partial J_\perp/\partial L)_E = 0$ for the interval 7—20 September 1966. Following a similar "injection" during the magnetic storm of 25 May 1967, the "crest" of a flux profile ($E > 0.5$ MeV) was observed to move such that $d L_c/dt \approx -0.02$/day between $L \approx 1.9$ and $L \approx 1.25$ [96].

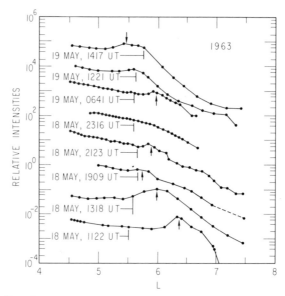

Fig. 57. Profiles of outer-zone electron flux ($E > 1.6$ MeV, omnidirectional) measured on Injun 3 during May 1963 [97]. Arrows denote instantaneous position of inward-moving "crest" (secondary maximum). Ordinate is separately normalized for each profile.

Typical "crest velocities" in the *outer* zone may be deduced from Fig. 57, which contains a selected interval of Injun-3 electron data ($E > 1.6$ MeV) obtained with a Type-302 Geiger tube [97]. Various secondary maxima of the integral flux appear to move inward with time, with a velocity $d L_c/dt \approx -10^{-2} (L/4)^8$ day^{-1}. This is about half the value typically deduced for a "leading-edge velocity" (see Fig. 52a). The dynamical significance of "velocities" obtained from such constant-energy data is somewhat dubious, although studies of this type have played an important historical role in radiation-belt phenomenology.

For the quantitative analysis of diffusion phenomena, natural electron "injection" events have the disadvantage of introducing spatially broad flux profiles, for which the competing dynamical processes are difficult to isolate. On the other hand, several of the artificial radiation belts created by high-altitude nuclear detonations were initially confined to rather narrow L ranges. Figure 58, for example, illustrates the evolution

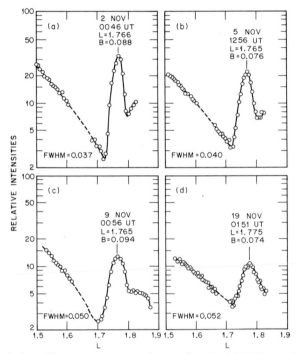

Fig. 58. Evolution of inner-zone electron-flux profile ($E > 1.9$ MeV, omnidirectional) observed on Explorer 15 following high-altitude nuclear explosion of 1 November 1962 [75].

of an electron-flux profile ($E > 1.9$ MeV) measured by an omnidirectional solid-state detector aboard the satellite Explorer 15 [75]. The narrow "spike" of electron flux centered at $L \approx 1.77$ had been injected by the Soviet high-altitude explosion of 1 November 1962. The full widths at half maximum (FWHM) were evaluated by fitting a Gaussian profile to each individual peak after subtracting the extrapolated background of Starfish-produced electrons (see Section IV.2). No systematic temporal shift in the position of the peak is apparent. The phenomenon of radial diffusion is evident, however, in the broadening of the flux profile with time. The value of $(\text{FWHM})^2$ appears to increase at a rate $\sim 7 \times 10^{-5}$

day^{-1}. If the indicated radial diffusion preserved particle energy rather than M and J, this numerical result could be translated[37] into a radial-diffusion coefficient $D_{LL} \sim 6 \times 10^{-6}$ day^{-1} [see (2.20)—(2.23), Section II.3].

The broadening of the profile contributed very little to the observed decrease in the trapped electron flux at $L \approx 1.77$. During the time span of the data plotted in Fig. 58, the maximum intensity decreased by a factor of approximately four, while the width increased by only $\sim 30\%$. Thus, a loss mechanism clearly must have been operating simultaneously with radial diffusion to account for the remaining decay (factor of 4/1.3) during the 17-day interval. The associated e-folding time (~ 15 days) is somewhat shorter than would have been expected from Fig. 41 (Section IV.3), in view of the fact that 1.9-MeV electrons in the outer zone are known to decay more *slowly* than 0.5-MeV electrons.

The 15-day lifetime deduced from Fig. 58 corresponds to $D_{xx} \approx 0.01$ day^{-1} (see Section II.7). If this pitch-angle diffusion coefficient is inserted in (3.40) or (3.45), however, the resulting radial-diffusion coefficient at $L \approx 1.77$ cannot exceed $\sim 8 \times 10^{-7}$ day^{-1} outside the loss cone (see Fig. 30, Section III.7). Shell splitting thus fails by a factor ~ 7 to account for the radial diffusion illustrated in Fig. 58. In the absence of other known radial-diffusion mechanisms that preserve E, it seems likely that the data should be analyzed at constant M and J (see Section V.6).

IV.7 Fluctuating Magnetospheric Fields

Compared with the extensive spatial and spectral measurements of the trapped electron and proton populations, the observation of wavelike electric- and magnetic-field fluctuations (plasma turbulence) in the magnetosphere remains in the basic exploratory stage. Of the wavelike phenomena that have been observed, most are related to pitch-angle diffusion rather than radial diffusion. The disturbances known to produce significant radial diffusion often extend coherently over the entire magnetosphere (see Sections III.2 and III.3) and so it may be inappropriate to characterize them as waves.

As indicated in Section II.5, electromagnetic waves propagating in the whistler mode are in the correct frequency band to cyclotron-resonate with radiation-belt electrons. All the various types of VLF signals that

---

[37]Division of the time derivative of (FWHM)^2 by 16ln 2 yields a radial diffusion coefficient compatible with (3.42). For a profile as narrow as that in Fig. 58, it is safe to neglect the term $L^{-2}[\partial(L^2 D_{LL})/\partial L](\partial \bar{f}/\partial L)_{E,x}$ in the expansion of (3.42).

propagate in the whistler mode (*e. g.*, chorus, hiss, hooks, and whistlers) can in principle produce pitch-angle diffusion, and it appears from satellite VLF data that only a small fraction of the measured VLF energy belongs to whistlers generated by lightning strokes (see Section II.3). Most of the wave energy is of magnetospheric origin.

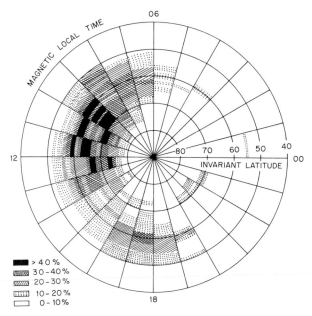

Fig. 59. Probability-of-occurrence distribution for wide-band VLF noise (0.2—7.0 kHz) with magnetic amplitude $> 1.8 m\gamma$ rms (root-mean-square), based on Injun-3 observations at altitudes up to 2700 km [98].

Calculations have shown that root-mean-square (rms) magnetic-field fluctuations $\sim 1 m\gamma$ in the frequency range of whistler-mode disturbances would be sufficient to account for the lifetimes shown in Fig. 41 for electrons that mirror at magnetic latitudes $\gtrsim 20°$ [43]. Noise of this magnitude is a common occurrence at altitudes up to ~ 2700 km in the invariant-latitude[38] range $50°—70°$. The probability-of-occurrence distribution for VLF signals $(\gtrsim 1.8 m\gamma)$ in the 0.2—7.0 kHz band, as measured on the Injun-3 satellite, is shown in Fig. 59 [98]. The VLF

[38]Low-altitude satellite data are frequently organized with respect to invariant magnetic latitude Λ, defined such that $L_m = \sec^2 \Lambda$. The invariant latitude of a field line is the magnetic latitude at which the line would intersect the earth's surface in a centered-dipole idealization.

noise drops off sharply below $\Lambda \sim 50°$ ($L \sim 2.5$) and is also largely absent in the quadrant centered on midnight. The drop-off in disturbance below $\Lambda \sim 50°$ may account for the reduced electron loss rates (longer lifetimes) found at the lower L values (see Fig. 41).

Measurements of energetic electrons (300 keV to 2.3 MeV) reveal a pronounced peak in the profile of precipitating outer-zone electrons [99]. The location of this peak is strongly correlated in L with the morningside location of the plasmapause (see Section I.6). Further, the morningside location of the plasmapause is observed to correlate well with the occurrence of ELF emissions. Apparently the most intense pitch-angle diffusion of electrons into the loss-cone ($x_b > x > x_c$) is that produced by the ELF emissions on the morning side of the magnetosphere. The electrons subsequently drift in longitude to the South Atlantic "anomaly" where they enter the atmosphere and are lost from the radiation belts.

In view of the coincidence in frequency, the Pc-4 ($\sim 10^{-2}$ Hz) micropulsation observed at ATS 1 and shown in Fig. 18d (Section II.3) may resonate with the bounce motion of ring-current protons (~ 15 keV). At synchronous altitude, the wave was seen only in the ζ (compressional) component. Spectral analyses of $\dot{\mathbf{B}}$ at ATS 1 and at College, Alaska, during this event are shown in Fig. 60 [100]. At College (which lies near the foot of the ATS-1 field line) the oscillation was almost entirely noncompressional. The relative spectral intensities suggest a 2% efficiency of mode conversion from an apparent magnetosonic wave at the equator (see Section II.5) to a transverse Alfvén wave [see (2.51b) Section II.6] in the lower ionosphere.

For other disturbances, notably sudden impulses (*si*) and sudden commencements (*sc*), the magnetic-field perturbation seen on the ground is similar in magnitude to that seen in the magnetosphere at synchronous altitude[39]. The factor relating such field perturbations seen on the ground and at ATS 1 is often compatible with that suggested in Fig. 9 (from the induced-dipole effect, Section I.5). Thus, a sudden impulse in b [equation (1.45), Section I.5] ideally produces

$$\dot{\mathbf{B}} \approx \hat{\theta}(\dot{b}/b)(a/b)^3 [3 B_1 - 4 B_2 (r/b) \cos \varphi] \qquad (4.02\text{a})$$

at the satellite ($r = 6.6a$, $\theta = \pi/2$) and

$$\dot{\mathbf{B}} \approx \hat{\theta}(\dot{b}/b)(a/b)^3 (9 B_1/2) \sin \theta \qquad (4.02\text{b})$$

on the ground ($r = a$). The φ dependence of $\mathbf{B}$ on the ground should be negligible, according to (1.45). The addition of a conducting earth

---

[39] An *sc* is followed by the main phase of a magnetic storm, while an *si* is not.

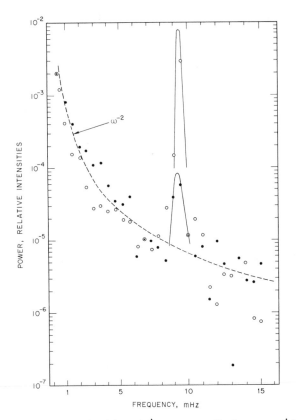

Fig. 60. Relative spectral densities of $\dot{\mathbf{B}}$, *i.e.*, contributions to $\langle\dot{\mathbf{B}}^2\rangle$ per unit frequency interval, at ATS 1 (open circles) and College (filled circles) during Pc-4 event shown in Fig. 18d [100]. Solid curves schematically trace central peak of each spectrum. Dashed curve represents inverse-square background common to both spectra.

[in (4.02b)] makes $\dot{B}_r$ vanish at $r=a$. In this idealization, temporal variations of **B** are excluded (on a sudden-impulse time scale) from the region $r<a$.

The magnetic spectral density $\mathscr{B}_z(\omega/2\pi)$ required in (3.13) is thus related to $\mathscr{B}_\theta(\omega/2\pi)$, as observed on the ground, by the simple transformation

$$\mathscr{B}_\theta(\omega/2\pi)\approx(9/4)\sin^2\theta\,\mathscr{B}_z(\omega/2\pi). \qquad (4.03)$$

If extraneous disturbances seen on the ground but not at the satellite (*e.g.*, resulting from currents localized in the ionosphere) are excluded from the spectrum by a careful search of the magnetograms for *si*'s and *sc*'s, it becomes possible to estimate the magnitude of D_{LL} due to the occurrence of these *si*'s and *sc*'s. By analogy with (3.20), the ground-based observations yield a spectral density

$$\mathscr{B}_\theta(\omega/2\,\pi) = \frac{2(\tau_d^2/\tau)\Sigma(\varDelta B_\theta)^2}{1+\omega^2\,\tau_d^2}, \tag{4.04}$$

which reduces to $\mathscr{B}_\theta(\omega/2\pi) \approx (2/\omega^2\,\tau)\Sigma(\varDelta B_\theta)^2$ in the limit $\omega^2\,\tau_d^2 \gg 1$, *i. e.*, for drift periods short compared to the characteristic decay time of an impulse.

A four-year compilation of equivalent equatorial ($\theta = \pi/2$) data obtained from ground-based measurements is summarized in Table 10 [101]. If the probability of *sc* and *si* occurrence (per year) is multiplied by the product of the impulse-amplitude bin limits, the sum over all amplitude bins yields the estimate that $(1/\tau)\Sigma(\varDelta B_\theta)^2 \approx 5.11 \times 10^4\gamma^2/\text{yr} \approx 140\gamma^2/\text{day}$. For $\omega^2\tau_d^2 \gg 1$, the *sc*'s and *si*'s thus apparently produce a radial-diffusion coefficient [40]

$$D_{LL} \approx 10^{-8}(a/b)^2 L^{10}\,\text{day}^{-1} \approx 10^{-10} L^{10}\,\text{day}^{-1} \tag{4.05}$$

for $b = 10a$ and $y = 1$ [see (3.14), Section III.2].

Table 10. Summary of Magnetic Impulses (1958—61)

Amplitude	Frequency	Contribution to $(10/L)^{10} D_{LL}$, day^{-1}	
$\varDelta B_\theta, \gamma$	Events/yr	Quasilinear	Nonlinear
> 100	0.5	0.17	0.25
60—99	1.8	0.20	0.31
40—60	2.3	0.10	0.12
20—40	21.	0.32	0.36
5—20	61.	0.16	0.16
~2	720.	0.05	0.05
Total	~ 800	1.00	1.25

Compiling a list such as Table 10 requires subjective judgements, and different investigators often disagree on the results. Values of D_{LL} ranging from 2—$4 \times 10^{-11} L^{10}\text{day}^{-1}$ (considering only *sc*'s [102]) up to 4—$13 \times 10^{-9} L^{10}\text{day}^{-1}$ (both *sc*'s and *si*'s [103]) have been reported [41].

---

[40]This is the "quasilinear" result. The "nonlinear" result (last column) is inferred from the published results of a nonlinear calculation [101] by requiring agreement with "quasilinear" theory for the smallest impulses.

[41]The estimate of 4—$13 \times 10^{-9} L^{10}\text{day}^{-1}$ was based upon a summary of sudden-impulse data from Explorer 12 as well as from ground-based records.

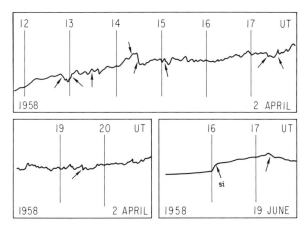

Fig. 61. Horizontal (H) component of magnetic records from Honolulu Observatory, showing sudden impulse (si) recognized in official compilations of such data [105] and similar fluctuations (unlabeled arrows) not present on enough other magnetograms in the global network to qualify as recognized sudden impulses [104].

Values up to $2 \times 10^{-10} L^{10} \, \text{day}^{-1}$ can be obtained in place of (4.05) by slightly modifying the manipulation of the data of Table 10 [101].

Part of the subjectivity involved in the inspection of ground-based magnetometer records is illustrated in Fig. 61 [104]. Several features (designated by arrows) certainly look like sudden impulses, but only one (that labeled si) is officially recognized as such in compilations of geomagnetic and solar data [105]. In the absence of any apparent morphological distinctions, it would seem that all events in Fig. 61 should be included in (4.04). This apparently has not been done in Table 10, which lists only about 800 events per year. Moreover, many of the events designated by arrows in Fig. 61 have rise times τ_r and decay times τ_d that do not fit the pattern $\tau_r \ll 2\pi/\Omega_3 \ll \tau_d$, where $2\pi/\Omega_3$ is the azimuthal-drift period of a radiation-belt particle. Such impulses will alter the functional form of $\mathscr{B}_0(\omega/2\pi)$, causing it to deviate (at high and low frequencies) from proportionality to ω^{-2}. The radial-diffusion coefficient is accordingly reduced for particles having $\Omega_3^2 \tau_r^2 \gtrsim 1$, just as (4.04) suggests a reduction from (4.05) for $\Omega_3^2 \tau_d^2 \lesssim 1$. In evaluating expressions for D_{LL} in such cases, it is convenient to recall [cf. (1.35), Section I.4] that

$$|2\pi/\Omega_3| = (m_e/m_0)(\gamma/L)(\gamma^2 - 1)^{-1} \times 173.2 \, \text{min} \qquad (4.06)$$

at $y = 1$ in the dipole field ($173.2 \approx 100\sqrt{3}$).

One way to avoid the subjectivity of visually identifying sc's and si's is to generate, directly from digitized magnetograms, a power spectrum of the field variations. The result of such an analysis of data obtained on the ground at $L \approx 2.7$ is shown in Fig. 62 [106] and corresponds to an equatorial spectrum

$$\mathscr{B}_\theta(\omega/2\pi) \approx 1.0 \times 10^{-2} [(2\pi/\omega) \div 1\,\mathrm{sec}]^2 \gamma^2/\mathrm{Hz} \qquad (4.07)$$

at frequencies $\omega/2\pi \lesssim 40\,\mathrm{mHz}$. According to (4.03) and (3.14), the spectrum yields a radial-diffusion coefficient

$$D_{LL} \approx 1.2 \times 10^{-8} L^{10}\,\mathrm{day}^{-1} \qquad (4.08)$$

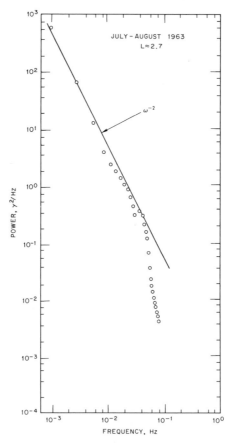

Fig. 62. Spectral density (1—200 mHz) of fluctuations in total field intensity B observed at Lebanon State Forest, New Jersey. Compilation includes all available data, regardless of local time or K_p index [106].

for equatorially mirroring particles (assuming $b = 10a$). This number is compatible with the maximum value ($\sim 13 \times 10^{-9} L^{10} \mathrm{day}^{-1}$, see above) reported by an investigator using sc's and si's, but is slightly large compared to the *total* D_{LL} that outer-zone particle observations appear to require (see Chapter V). The result given by (4.08) is ~ 100 times that given by (4.05) for what is alleged to be the same diffusion mechanism, viz., the violation of Φ by sudden *magnetic* perturbations.

Part of the disagreement[42] that is noted among results differently obtained [cf. (4.05) and following discussion] may arise from the production of ionospheric currents by impulses that are *electrostatic* in the magnetosphere. These currents would cause **B**-field perturbations that are detectable on the ground (cf. Fig. 61) but insignificant in the outer zone. Another possibility is that magnetic-field perturbations in space are magnified by induced ionospheric currents to yield an enhancement on the ground that far exceeds the factor $\sim 3/2$ implied by (4.02). Such an effect may help to account for the fact that the *diurnal* variation of **B** at the earth's surface during quiet periods is ~ 10 times the amplitude $\sim 2\gamma$ predicted [28] by (1.45).

According to (4.02) a sudden magnetic impulse should have a nearly φ-independent amplitude at the ground, but a strongly φ-dependent amplitude (larger at noon than at midnight) in the vicinity of synchronous altitude. Comparisons of sudden-impulse amplitudes observed at ATS 1 and on the ground at Honolulu (cf. Section III.1) sometimes confirm this expected diurnal variation of their ratio, while at other times such comparisons are inconclusive [107]. As suggested above, the ionosphere (which is, on the average, a much better conductor by day than at night) may play a role that is not yet understood.

The results indicated by (4.05) and (4.08) are thus subject to the considerable uncertainties that often surround the quantitative application of ground-based observations to magnetospheric phenomena. Table 10 covers a period of moderate solar and geomagnetic activity, and the consequent D_{LL} given by (4.05) would perhaps have been larger at solar maximum and smaller at solar minimum (cf. Fig. 51, Section IV.5). The data for Fig. 62 cover only a few selected days in 1963 and do not represent a true average of any sort. It is difficult in any case to reconcile the disparate values of D_{LL} indicated by (4.05) and (4.08).

For several of the larger impulses listed in Table 10, the "quasilinear" analysis leading to (3.13) could easily be inadequate. A pair of radiation belt particles initially coincident in L (~ 5) but $180°$ out of phase in

---

[42]Aside from obvious considerations, such as the fact that the various data were acquired during different time periods.

drift could easily find themselves separated by $\Delta L \gtrsim 1$ following an impulse $\gtrsim 50\gamma$ in amplitude [*cf.* (3.11)]. Moreover, a random succession of such large impulses may violate (3.13) to some extent, by virtue of a nonlinearity in the relationship among L, dL/dt, and db/dt [*cf.* (3.12), Section III.2]. As a consequence of this nonlinearity, the radial-diffusion coefficient implied by the data of Table 10 is probably somewhat larger than (4.05) indicates, and the relative contribution of the larger impulses is somewhat enhanced (see Table 10, last column) [101].

IV.8 Drift Echoes

As noted in Section III.1, magnetic sudden impulses are frequently followed by drift-periodic echoes in the outer-zone particle fluxes. Figure 23 (Section III.1) illustrated this phenomenon for unidirectional ($\approx J_\perp$) electron fluxes at synchronous altitude. In Fig. 63, drift echoes are observed in the fluxes measured by the same seven unidirectional electron channels on the ATS-1 satellite [108]. The characteristic echo frequency observed in each distinctly echoing channel of Fig. 63 is plotted (Fig. 64) against the value of $(\gamma^2 - 1)/\gamma$ corresponding to the respective nominal electron energy. The resulting linear relationship (with a reasonable

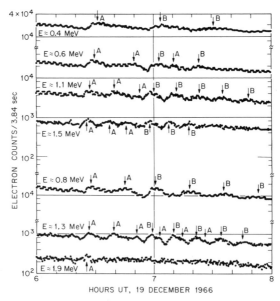

Fig. 63. Unidirectional electron fluxes (roughly $J_\perp$ channels) measured on ATS 1 during period of overlapping drift-echo events, denoted *A* and *B* [108].

magnitude for Ω_3, as shown below) confirms the drift-periodicity of the echoes[43]. If $J_\perp$ is proportional to E^{-l} on the drift shell in question (*cf.* Section II.6), the typical spread in Ω_3 may be estimated as $\Delta\Omega_3 \sim \Omega_3/l$ for a given broad energy channel. The drift-echo amplitude therefore decays with an *e*-folding time $\sim l/\Omega_3$ ($\sim l/2\pi$ drift periods) by virtue of phase mixing in φ_3 among particles detected in the same energy channel (see Sections II.1 and III.1).

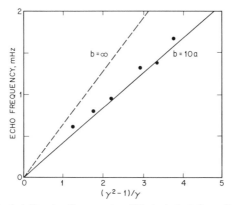

Fig. 64. Empirical drift-echo frequencies (filled circles) from Fig. 63 for nominal electron energies $E=(\gamma-1)m_0 c^2$. Theoretically expected drift frequencies for pure-dipole field (dashed line) and compressed-dipole field (solid line) are derived from (4.09).

The magnitude of Ω_3 in Fig. 64 is smaller than would be expected in a dipole field ($b=\infty$) for the same electron energy and drift-shell radius. For equatorially mirroring particles the magnetospheric compression represented by B_1 in (1.45) acts to reduce the angular drift velocity. If B_2 is neglected altogether in (1.45), the equatorially mirroring particle sees a guiding-center force [*cf.* (3.32)] equal to $-(M/\gamma)(\partial B/\partial r)$ in the radial direction. The resulting azimuthal drift is such that

$$\Omega_3 = v_d/r = (c/qBr)(M/\gamma)(\partial B/\partial r)$$
$$= -3B_0(m_0 c/2q B^2 \gamma)(\gamma^2-1)(c/a)^2(a/r)^5, \qquad (4.09)$$

---

[43]The dependence of Ω_3 on y is very weak (*cf.* Fig. 7) and can be neglected here, even if the form of the pitch-angle distribution varies with electron energy. The threshold energy is regarded as the nominal energy of particles that count in a given channel. If the energy spectrum were oddly shaped, or the threshold energy difficult to identify from the detector counting-efficiency curves $\varepsilon_j(E)$, it might be necessary to plot the function $J_\perp(E)(dE/d\Omega_3)\varepsilon_j(E)$ against $(\gamma^2-1)/\gamma$ for each energy channel (j) in order to identify the mean (or most probable) drift frequency and relative bandwidth.

where $B=B_0(a/r)^3+B_1(a/b)^3$. The drift frequency expected from (4.09) is found to agree well with that observed in each energy channel (Fig. 64). Moreover, equation (4.09) clearly predicts that $(\partial\Omega_3/\partial r)_E=0$ at $r=(B_0/5B_1)^{1/3}b\approx(\pi/5)b\approx6.3\,a$, i. e., at $B=6B_1(a/b)^3\approx150\gamma$, if $b=10\,a$. Thus, the drift frequency of a particle having $y=1$ and a fixed energy attains its maximum with respect to r at a location $\sim0.3\,a$ below synchronous altitude[44].

Drift echoes are produced because a sudden impulse differentially (with respect to φ_3) changes the energy and L value of a population of trapped particles. In other words, the magnitudes of ΔE and ΔL depend upon a particle's drift phase at the instant of the impulse (see Fig. 65) [55]. The effect of magnetic impulses on a particle population having $J=0$ can be estimated by referring to the model in Section III.2.

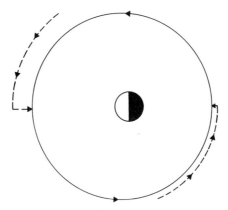

Fig. 65. Schematic illustration of third-invariant violation by sudden compression of magnetosphere. Electrons initially on different drift paths (dashed curves) are accidentally $180°$ apart in drift phase at time of compression that puts them both onto the same, new drift shell. The mean inward displacement and energization is adiabatic, related to the global increase in $B_1(a/b)^3$. The particle energized on the day side has thus moved inward in L, while the particle less energized on the night side has moved outward in L, as defined by (1.37).

The equatorial $\mathbf{B}$ field seen by a particle in the magnetic-field model specified by (1.45) points in the $-\hat{\boldsymbol{\theta}}(=+\hat{\mathbf{z}})$ direction and has a magnitude

$$B_e(r,\varphi;t)=B_0(a/r)^3+B_1(a/b)^3-B_2(a/b)^3(r/b)\cos\varphi\,. \qquad (4.10)$$

---

[44]The accuracy of (4.09) is perturbed only to *second* order in B_2 if the azimuthal asymmetry of $\mathbf{B}$ is included in the calculation of Ω_3.

According to Table 7 (Section III.2), a magnetic impulse induces an electric field

$$E_\varphi(r, \varphi; t) \approx (3/2\,c)(a/b)^3\,(d\,b/d\,t)\{B_1(r/b)$$
$$- (64/63)\,B_2(r/b)^2\,[1 + (13\,B_1/832\,B_0)(r/b)^3\cos\varphi] \qquad (4.11\,\text{a})$$
$$+ (2\,B_2/63\,B_0)\,B_2(r/b)^6\cos 2\,\varphi\}$$

$$E_r(r, \varphi; t) \approx - (4/7\,c)(a/b)^3\,(d\,b/d\,t)\{B_2(r/b)^2\,[1 - (9\,B_1/52\,B_0)(r/b)^3]\sin\varphi$$
$$- (7\,B_2/25\,B_0)\,B_2(r/b)^6\sin 2\,\varphi\} \qquad (4.11\,\text{b})$$

in the equatorial plane $(\theta = \pi/2)$ of this magnetic-field model.

For $db/dt = 0$, an equatorially mirroring particle would follow a path of constant B, as given by (4.10). A sudden impulse (at time $t = t_0$) alters the value of B seen by the particle; the rate of change is given by

$$d\,B_e/d\,t = (\partial\,B_e/\partial\,t) + c(E_\varphi/B)(\partial\,B_e/\partial\,r) - c(E_r/r\,B_e)(\partial\,B_e/\partial\,\varphi)$$
$$\approx - (3/b)(a/b)^3\,(d\,b/d\,t)[(5/2)\,B_1 - (20/7)\,B_2(r/b)\cos\varphi] \qquad (4.12)$$

to lowest order in ε_1 and ε_2 (see Section I.7). The resulting phase organization can be traced via Liouville's theorem (Section I.3). It can thus be shown that

$$\ln f(M, J, L_m(t_0^+), \varphi_3; t_0^+) - \ln \overline{f}(M, J, L_m(t_0^-); t_0^-)$$
$$\approx \{1 - (1/2\,\gamma)(\gamma + 1)(\partial \ln \overline{J}_\perp/\partial \ln E)_L \qquad (4.13)$$
$$+ (1/3)(\partial \ln \overline{J}_\perp/\partial \ln L_m)_E\} \ln[B_e(t_0^+)/B_e(t_0^-, \varphi_3)]$$

for $J = 0$, where $B_e = B_0/L_m^3$ (see Section I.5). For an impulse with vanishing rise time and small amplitude, it follows from (4.12) that

$$\ln[B_e(t_0^+)/B_e(t_0^-, \varphi_3)]$$
$$\approx (5\,L_m^3/2\,B_0)[1 - (8\,B_2/7\,B_1)(L_m\,a/b)\cos\varphi_3]\,\Delta[B_1(a/b)^3], \quad (4.14)$$

where the drift phase φ_3 is defined in terms of a particle's azimuthal coordinate φ at time $t = t_0$ (see Section I.5). Following the sudden impulse, as **B** remains constant in time, the evolution of $J_\perp \equiv p^2 f$ is given by

$$\ln J_\perp(E, x = 0, L_m, \varphi; t) - \langle \ln J_\perp(E, x = 0, L_m, \varphi_3)\rangle$$
$$\approx \{1 - [(\gamma + 1)/2\,\gamma](\partial \ln \overline{J}_\perp/\partial \ln E)_L + (1/3)(\partial \ln \overline{J}_\perp/\partial \ln L_m)_E\} \qquad (4.15)$$
$$\times (- 20\,L_m^4\,B_2/7\,B_1\,B_0)(a/b)\,\Delta[B_1(a/b)^3]\cos[\varphi - \Omega_3(t - t_0)],$$

where the angle brackets denote the drift average for $t > t_0$. Thus, as a result of the impulse at $t = t_0$, the spatial and spectral structure of the previously ($t < t_0$) "unperturbed" magnetosphere unfolds in *time* as the representative particles subsequently drift past the satellite.

According to (4.15), the drift echoes observed at local noon ($\varphi = \pi$) will tend to have a smaller amplitude than those seen at local midnight ($\varphi = 0$) because of the factor L_m^4. It follows that more events will escape detection when the satellite is at noon than when it is at midnight. A compilation of drift-echo events by probability of occurrence (Fig. 66) shows a diurnal variation having such an interpretation [109]. Here electron drift-echo events have been counted subject to the requirement that $\ln J_\perp$ "oscillates" with a peak-to-peak amplitude $\gtrsim 1/2$. Using a finer "resolution", $(\Delta \ln J_\perp)_{pp} \gtrsim 1/10$, it is possible to identify up to ~ 20 drift-echo events per day during periods of moderate geomagnetic activity.

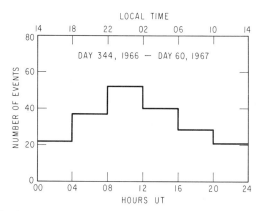

Fig. 66. Distribution of 200 major drift-echo events identified in ATS-1 electron data during an 82-day interval, in terms of satellite local time at beginning of event [109].

The compilation shown in Fig. 66 can be utilized, in principle, to estimate D_{LL} at synchronous altitude. If B_2 is tentatively neglected in order to simplify the field model [as in (4.09)] it can be shown without further approximation that

$$L \equiv 2\pi a^2 B_0 |\Phi|^{-1} = (r/a)[1 - (B_1/2 B_0)(r/b)^3]^{-1} \tag{4.16a}$$

$$L_m \equiv (B_0/B_e)^{1/3} = (r/a)[1 + (B_1/B_0)(r/b)^3]^{-1/3}, \tag{4.16b}$$

where r is the equatorial drift-shell radius. At $r = 6.6a$ it follows that $L = 7.37$, $L_m = 6.16$, and $dL/dL_m = 1.93$. The radial-diffusion coefficient is thus given by

$$D_{LL} = (dL/dL_m)^2 (L_m^2/2\,\tau)\Sigma\langle(\Delta \ln L_m)^2\rangle$$
$$= (dL/dL_m)^2 (L_m^2/18\,\tau)\Sigma\langle(\Delta \ln B_e)^2\rangle$$
$$= (dL/dL_m)^2 (L_m^2/144\,\tau)\Sigma(\Delta \ln B_e)_{pp}^2 \qquad (4.17)$$
$$= (dL/dL_m)^2 (L_m^2/144\,\tau)\Sigma(\Delta \ln J_\perp)_{pp}^2$$
$$\div \{1 - [(\gamma+1)/2\,\gamma](\partial \ln \bar{J}_\perp/\partial \ln E)_L - (\partial \ln \bar{J}_\perp/\partial \ln B_e)_E\}^2 ,$$

where pp denotes the peak-to-peak amplitude.

The term $-(\partial \ln \bar{J}_\perp/\partial \ln B_e)_E$ in (4.17) is identical with the term $+(1/3)(\partial \ln \bar{J}_\perp/\partial \ln L_m)_E$ in (4.13) and (4.15). It can be estimated as ranging from ~ 2 at $E=0.4\,\mathrm{MeV}$ to ~ 6 at $E=1.5\,\mathrm{MeV}$ by measuring the diurnal variation of the electron fluxes at synchronous altitude on a sufficiently quiet day. A synchronous satellite measures a smaller flux in each *energy* channel at midnight than at noon (see Fig. 50, Section IV.5) because L_m is larger at midnight than at noon [6.39 *vs* 5.95, according to (4.10)]. The spectral index $-(\partial \ln \bar{J}_\perp/\partial \ln E)_L$ also shows a variation with energy for outer-zone electrons (~ 1 at $E=0.4\,\mathrm{MeV}$ to ~ 5 at $E=1.5\,\mathrm{MeV}$). The divisor of $(\Delta \ln J_\perp)_{pp}^2$ in (4.17) is therefore of order unity[45], indicating that $\Delta \ln J_\perp$ (as seen by the observer) is roughly equal to $\Delta \ln B_e$ (as seen by a particle with $\varphi_3 \sim \pi$). According to (4.12), the value of dB_e/dt is 4.6 times as large for $\varphi_3 = \pi$ as for $\varphi_3 = 0$; strictly speaking, $(\Delta \ln B_e)_{pp}$ is the *difference* between the change in $\ln B_e$ experienced by a particle at noon and that seen by a particle at midnight.

Figure 66 indicates an average of two drift-echo events per day for which $(\Delta \ln J_\perp)_{pp} \gtrsim 1/2$. As some events have amplitudes larger than the nominal threshold, the estimate that $(1/\tau)\Sigma(\Delta \ln J_\perp)_{pp}^2 \sim 1\,\mathrm{day}^{-1}$ appears reasonable. It then follows from (4.17) that $D_{LL} \sim 1\,\mathrm{day}^{-1}$ $\sim 10^{-9} L^{10}\,\mathrm{day}^{-1}$ at synchronous altitude (using $L=7.37$, as noted above). This result is in line with diffusion coefficients extracted from other data (*cf.* Section IV.7 and Chapter V). The result can be extrapolated to $L \lesssim 4$ by using (4.10) and (4.11) in full, *i.e.*, retaining terms of order ε_1^2, ε_2^2, $\varepsilon_1\varepsilon_2$, and so on. The diffusion coefficient thus calculated for magnetic sudden impulses has the property that $(1/2\,\tau)\langle(\Delta L_m)^2\rangle$ at $L_m = 6.16$ is approximately $(2/3)(6.16)^{10}$ times the value of D_{LL} at $L_m=1$. Since $(dL/dL_m)^2 \approx 15/4$ at $L_m = 6.16$, it follows that

$$D_{LL} \sim (3/2)(4/15)(L/6.16)^{10} \times 1\,\mathrm{day}^{-1}$$
$$\sim 5 \times 10^{-9} L^{10}\,\mathrm{day}^{-1} \qquad (4.18)$$

---

[45]The factor $\{1 - [(\gamma+1)/2\gamma](\partial \ln \bar{J}_\perp/\partial \ln E)_L - (\partial \ln \bar{J}_\perp/\partial \ln B_e)_E\}$ apparently was *positive* on 22 February 1967 (Fig. 23). The data show a minimum in the flux at $\varphi_3 \sim \pi$ and a maximum at $\varphi_3 \sim 0$ after a sudden decrease of B_e.

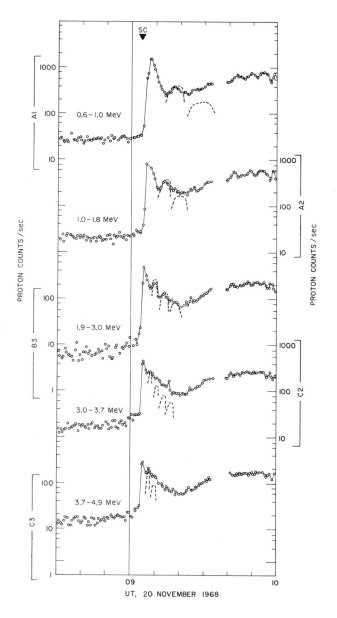

Fig. 67. Evidence of proton drift echoes initiated by sudden commencement at 0904 UT and observed at ATS 1 near local midnight. Dashed curves represent predicted flux modulation based on shape of initial pulse, energy spectrum, and detector response function for each unidirectional channel, assuming a Mead-Williams magnetosphere with $b=9a$ and $B_t=40\gamma$ [110].

at $L \lesssim 4$. It is important, however, to recognize that the parameters inserted in (4.17) to obtain this result are subject to serious uncertainties, and that confidence in the numerical magnitude of this D_{LL} must be tempered accordingly.

Drift echoes have also been detected (following a sudden commencement) in the *proton* flux at ATS 1. As discussed in Section IV.3, proton enhancements and their subsequent decay are frequently observed at synchronous altitude following a sudden change in the earth's magnetic field. The proton data plotted in Fig. 67 show that one or two additional enhancements of the flux follow the initial (large) enhancement that coincides with the sudden commencement [110]. The dashed curves on the figure are the predicted proton intensities, according to a model that the initial flux enhancement is localized in a 30° sector of longitude. The proton spectrum in the model is taken to be the proton spectrum at the peak of the enhancement. As the protons subsequently drift around the earth, they disperse in drift phase. Since the satellite instrument detects protons that vary in energy over a finite "window", the echo of the *sc*-associated enhancement disperses with time. By computer simulation of the detector response, it is found that satisfactory agreement with the observations can be achieved by using a Mead-Williams magnetosphere (see Section I.5) with $b = 9a$ and $B_t = 40\gamma$.

The protons in Fig. 67 are the solar-flare protons which have relatively free access to the outer regions of the magnetosphere (Section IV.3). Since they undoubtedly experience radial diffusion, it is possible for the magnetosphere to trap some of them. These captured solar-flare protons may ultimately diffuse radially (at constant M and J) into the inner magnetosphere where they would contribute to the high-energy proton population.

V. Methods of Empirical Analysis

V.1 Basic Objectives

After obtaining experimental data such as those discussed in the previous chapter, it is necessary to extract from the data numerical values of transport coefficients such as D_{xx} and D_{LL}, in order best to describe the observations in the context of magnetospheric diffusion processes. This must be done with the realization that such a course of action (and the parameters determined from it) are subject to uncertainties, including the question as to whether the observations can actually be described in terms of the diffusion processes selected.

At present, the most desirable course to take in verifying the validity of the diffusion equation adopted for describing a given set of magnetospheric particle observations is to use a self-consistent analysis. In a self-consistent approach, the values of the transport coefficients should be determined (as far as possible) empirically from the particle data. After values for the coefficients are so estimated, the model should be verified by inserting these values in the appropriate diffusion equation to show that the model indeed predicts the observed spatial structure and/or temporal evolution of the particle data. Subsequently, the magnitudes of the transport coefficients should be manipulated (by the methods of Chapters II and III) to yield predictions for the spectral densities of magnetospheric field fluctuations (waves, impulses, *etc.*). Finally, these predictions should be compared with available observations of magnetospheric field and wave activity (see Section IV.7).

An empirical determination of the transport coefficients directly from the measured particle data presents certain difficulties. Frequently, such a determination is arrived at by making initial assumptions concerning the relative importance of the various diffusion mechanisms. Often the transport coefficient associated with the dominant process can be determined only by assigning to the other (secondary) process a fixed and somewhat arbitrary value. Moreover, it is usually necessary to assume that the transport coefficients are time-independent, or else related in some fixed way to the geomagnetic indices (K_p, D_{st}, *etc.*).

For example, a common procedure not readily justified is that of directly relating the *apparent* electron-flux decay rates to numerical values of the pitch-angle diffusion coefficient D_{xx}. This approach, in which it is assumed (*cf.* Section II.7) that $D_{xx} \sim -(x_c^2/5)(\partial \ln \bar{f}/\partial t)_{E,x}$, is based on the expectation that the electron pitch-angle distribution is in its lowest eigenmode. The "decay" times shown in Fig. 41 (Section IV.3) have been obtained under this assumption.

A difficulty in this approach is illustrated by the $L=4$ electron data ($E > 1.0$ MeV) plotted in Fig. 38. During the first several days following each of the four largest magnetic storms, the electron flux did not appear to decay at all, but rather remained constant or increased in intensity. Thus, it would not have been possible to read a pitch-angle diffusion coefficient directly from these data. Electron losses undoubtedly occurred during these periods, but the temporal flux changes were probably dominated by radial-diffusion effects (*cf.* Figs. 52 and 53, Section IV.6).

Additional opportunities for determining the pitch-angle diffusion coefficient are provided by the data showing the azimuthal variation of precipitating electron fluxes (Fig. 36, Section IV.2) and the data showing relaxation of electron pitch-angle distributions to their lowest eigenmode (Fig. 34, Section IV.2). In these cases it is impossible to read D_{xx} directly from the data, and so more sophisticated analytical techniques are required (see Section V.2). Application of such techniques may yield both a nominal value and functional form for D_{xx}.

In the case of radial diffusion, several techniques and procedures have been developed for extracting D_{LL} from the observations. The choice of method depends in part on whether the data provide stationary (*cf.* Section IV.5) or time-varying (Section IV.6) flux profiles. As noted in Section IV.6, it is very helpful to have data in several energy channels in order to characterize the actual particle spectrum. Such spectral information affords considerable freedom in the choice of method for extracting D_{LL}.

When the data consist solely of "stationary" flux profiles, it is generally necessary to assign either D_{xx} or D_{LL} somewhat arbitrarily in order to extract the other. In the event that the data are obtained from the region of the magnetosphere where atmospheric losses predominate over wave-particle scattering, it is appropriate to insert D_{xx} as a known function of M, J, and Φ [*cf.* (2.17), Section II.2]. Outside the region where atmospheric scattering losses predominate, it is usually necessary to assign the observed "lifetimes" (as from Fig. 41, Section IV.3) characteristic of geomagnetically more active time periods (when the flux profiles typically are not stationary). Then, treating the relevant diffusion equation (see Sections III.7 and III.8) as a linear first-order differential equation for D_{LL}, it is possible to express the solution as a *spatial quadrature*, i.e., an integral with respect to L (see Section V.3). For this purpose, the derivatives of $\bar{f}$ are obtained numerically from the observational data. The radial-diffusion coefficient D_{LL} thus extracted from the data can subsequently be verified by obtaining the steady-state solution of the diffusion equation, *i. e.*, by *spatially integrating* the equation (see Section V.7) for $\bar{f}(M,J,\Phi)$. If this solution reproduces the observed steady-state profiles, then the value obtained for D_{LL} is considered reliable. If not, then either D_{LL} must be adjusted to yield a better fit, or the underlying model of the competing processes (*e. g.*, the arbitrarily assigned value of D_{xx}) must be modified.

Additional opportunities for empirical analysis arise if the observed flux profiles vary with time. The temporal coordinate adds a new dimension to the problem and makes it possible (in principle) to extract *both* D_{LL} and D_{xx} (as functions of L) from the data simultaneously. The introduction of a *variational technique* (see Section V.5) serves this purpose well. Simpler techniques allow *either* D_{xx} or D_{LL} to be expressed in terms of *quadratures* (Section V.4) over the data if the other diffusion coefficient is specified *a priori*. The ultimate test of the numerical validity of D_{xx} and D_{LL} obtained by any method is provided by a comparison of the data with the time-dependent solution of the diffusion equation, as obtained by *temporal integration* (Section V.6). In the present chapter, these various analytical methods for extracting diffusion coefficients from the data are discussed in a somewhat logical sequence.

V.2 Pitch-Angle Eigenmodes

Pitch-angle diffusion at constant energy is governed by (2.73), an equation that can also be written in the form

$$\frac{\partial \bar{f}}{\partial t} = \frac{1}{x} \frac{\partial}{\partial x} \left[x D_{xx} \frac{\partial \bar{f}}{\partial x} \right]_E - \frac{x}{y} D_{xx} \frac{T'(y)}{T(y)} \left[\frac{\partial \bar{f}}{\partial x} \right]_E, \qquad (5.01)$$

since $x^2 + y^2 = 1$. The second term of (5.01) is negligible for $x^2 \ll 1$. The approximation of omitting it altogether by taking $T(y) \equiv T(1)$ converts (5.01) to a diffusion equation in cylindrical coordinates. The eigenfunctions of (5.01) for an x-independent diffusion coefficient D_{xx} would then be Bessel functions of order zero (*cf.* Section II.7). In terms of (3.51) this would mean that the typical eigenfunction $g_n(x)$ is given by $g_n(x) = [2/T(1)]^{1/2} [1/x_c J_1(\kappa_n)] J_0(\kappa_n x/x_c)$, where $J_0(\kappa_n) = 0$. The corresponding eigenvalues λ_n of (3.52b) would then be of the form $\lambda_n = (\kappa_n/x_c)^2 D_{xx}$.

Generally, a source[46] term must be added to (5.01) in order to describe the evolution of $\bar{f}$ toward a steady-state distribution. If an

---

[46]In this section, the source can be regarded either as a true source or as a simulation of the radial-diffusion term omitted from (5.01).

isotropic source term is added to (5.01) [cf. (3.57a), Section III.8], then the pitch-angle distribution function $\bar{f}(x,t)$ will evolve in time toward the steady-state solution

$$\bar{f}_\infty(x) = (S x_c^2 / 4 D_{xx})[1 - (x/x_c)^2] \tag{5.02}$$

for $T(y) \equiv T(1)$ and D_{xx} independent of x. Equation (5.02) is reminiscent of (2.64) for $s = 1$ (see Section II.6).

The 28 October 1962 nuclear blast had injected electrons with an off-equatorial maximum that subsequently decayed with time (Fig. 34, Section IV.2). The decay of the omnidirectional flux ($E > 1.9$ MeV) can be simulated by expressing the pitch-angle distribution $\bar{f}(x,t)$ in terms of its steady-state solution and higher-order eigenmodes, i.e.,

$$\bar{f}(x, t) = \bar{f}_\infty(x) + \bar{f}_\infty(0) \sum_n a_n(t) J_0(\kappa_n x/x_c) \tag{5.03}$$

with $a_0(0) = 50/3$ and $a_1(0) = -a_2(0) = -55/3$. The artificial enhancement (Day 301.3) corresponds to $t = 0$, and $a_n(0)$ is assumed to vanish for $n > 2$.

The temporal decay of $\bar{f}(x,t)$ to $\bar{f}_\infty(x)$ obtained from (5.03) by assuming $x_c = 0.94$ and $D_{xx} = 10^{-2}$ day^{-1} is indicated in Fig. 68a [43]. The eigenvalues $\lambda_n = (\kappa_n/x_c)^2 D_{xx}$ are given by $\lambda_0 = 0.190 \lambda_1 = 0.077 \lambda_2 = 6.545 \times 10^{-2}$ day^{-1}. The differential omnidirectional flux at any point on the field line is given [cf. (1.22), Section I.4] by

$$J_{4\pi}(X, t) = 4\pi p^2 \int_0^1 \bar{f}(x, t) d(\cos\alpha)$$

$$= 4\pi p^2 \int_X^{x_c} \left[\frac{1 - X^2}{x^2 - X^2}\right]^{1/2} \frac{x \bar{f}(x, t)}{1 - X^2} dx, \tag{5.04a}$$

where

$$X^2 \equiv 1 - (B_e/B) = (x^2 - \cos^2\alpha)\csc^2\alpha. \tag{5.04b}$$

The coordinate X locates a point on any field line in terms of the local field intensity B relative to the minimum (or equatorial) field intensity B_e on that field line (cf. Section I.4). If D_{xx} and x_c are independent of energy (as assumed above), then pitch-angle diffusion leaves the form of the energy spectrum used in (5.03) invariant. The integral omnidirectional flux $I_{4\pi}(X,t)$ will then scale as $J_{4\pi}(X,t)$, which is plotted in Fig. 68b. The predicted evolution of $I_{4\pi}(X,t)$ thus resembles the observed evolution (Fig. 34) rather closely.

In both the observed and predicted $I_{4\pi}(X,t)$, the off-equatorial ($X > 0$) peak disappears as the pitch-angle distribution approaches its lowest decaying eigenmode. From $t \sim 10$ days onward, there is very little change in the form of the pitch-angle distribution (Fig. 68a). This means only

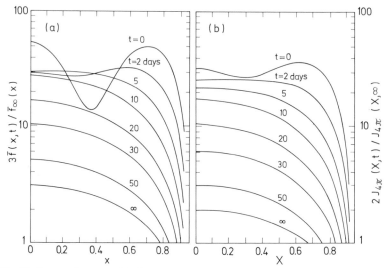

Fig. 68. Decay of equatorial pitch-angle distribution (a) and off-equatorial distribution of omnidirectional flux (b) to steady state (*cf.* Fig. 34), assuming $D_{xx} = 10^{-2}$ day^{-1} and $x_c = 0.94$ [43].

that the lowest eigenfunction, namely $J_0(\kappa_0 x/x_c)$, qualitatively resembles $1 - (x/x_c)^2$, which contains the entire pitch-angle dependence of $\bar{f}_\infty(x)$ for an isotropic source S [*cf.* (5.02)] if D_{xx} is independent of x.

The resemblance between $g_0(x)$ and $\bar{f}_\infty(x)$ exists even when D_{xx} varies with x. For example, if D_{xx} is proportional to $(x/x_c)^{2\sigma}$ with $\sigma < 1$, it is possible to integrate the equation [*cf.* (5.01)]

$$\frac{1}{x}\frac{d}{dx}\left[xD_{xx}\frac{d\bar{f}_\infty}{dx}\right] + S = 0, \qquad (5.05\,\text{a})$$

with the boundary condition that $\bar{f}_\infty(x_c) = 0$, so as to obtain

$$\bar{f}_\infty(x) = [Sx_c^2/4(1-\sigma)D_{xx}](x/x_c)^{2\sigma}[1 - (x/x_c)^{2-2\sigma}] \qquad (5.05\,\text{b})$$

under the assumption that $\partial S/\partial x = 0$. On the other hand, the normalized eigenfunctions of (3.52b) and (3.53) for $T(y) \equiv T(1)$ are given by

$$g_n(x) = -[2(1-\sigma)/T(1)]^{1/2}[1/x_c J_\nu'(\kappa_{\nu n})]$$
$$\times (x_c/x)^\sigma J_\nu(\kappa_{\nu n} x^{1-\sigma}/x_c^{1-\sigma}) \qquad (5.06\,\text{a})$$

where $\nu \equiv \sigma/(1-\sigma)$ and $J_\nu(\kappa_{\nu n}) = 0$ $(n = 0, 1, 2, \ldots)$. The corresponding eigenvalues are given by

$$\lambda_n = (1-\sigma)^2(\kappa_{\nu n}/x_c)^2(x_c/x)^{2\sigma}D_{xx}. \qquad (5.06\,\text{b})$$

Recall that $(x_c/x)^{2\sigma}D_{xx}$ is independent of x both in (5.06) and in (5.05). As shown in Fig. 69, where $x_c=0.9$, there is a close qualitative resemblance between $g_0(x)/g_0(0)$ and $\bar{f}_\infty(x)/\bar{f}_\infty(0)$ for $|\sigma|\lesssim 1/4$. Thus, the form of the pitch-angle distribution changes very little between the exponential-decay phase (see Fig. 34, Section IV.2) and the steady state.

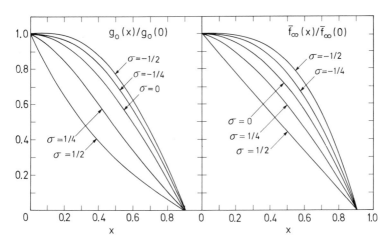

Fig. 69. Lowest eigenmode (left panel) and asymptotic steady state (right panel) for pitch-angle diffusion with $D_{xx}\propto(x/x_c)^{2\sigma}$ and $x_c=0.9$; steady state assumes isotropic source [43].

For $\sigma>0$ the pitch-angle diffusion coefficient vanishes at $x=0$. Consequently, the functions $g_0(x)$ and $\bar{f}_\infty(x)$ are more sharply peaked at $x=0$ for $\sigma>0$ than for $\sigma=0$. Conversely, if $\sigma<0$ the pitch angle distribution tends toward a broader shape, hence a steeper gradient at $x=x_c$. In either case, the value of D_{xx} at $x=x_c$ remains finite. Moreover, the values of $(1-\sigma)\kappa_{vn}$ that appear in (5.06b) are only moderately sensitive[47] to σ. The decay rates that govern the evolution of $\bar{f}(x,t)$ from $t=0$ to $t=\infty$ are thus largely insensitive to the manner in which pitch-angle diffusion is distributed over x.

The approximation that $T(y)\equiv T(1)$ in (5.01) causes the second term in that equation to vanish. This approximation is inappropriate for describing the temporal evolution of $\bar{f}(x,t)$ near $x\approx 1$ because the exact eigenfunctions of (5.01) are poorly approximated by (5.06a) for $x\approx 1$.

---

[47]For example, the quantity $(1-\sigma)\kappa_{v0}$ varies from 3.14 to 1.92 as σ goes from -1 to $+1/2$. The ratio λ_1/λ_0 varies from 9.0 to 3.4 over this same σ interval.

Without the approximation, the eigenvalue equation becomes [cf. (3.52b), Section III.8]

$$\lambda_n = -\frac{1}{x\,g_n}\frac{\partial}{\partial x}\left[x D_{xx}\frac{\partial g_n}{\partial x}\right]_E + \frac{D_{xx}}{y}\frac{T'(y)}{T(y)}\left[\frac{\partial \ln g_n}{\partial \ln x}\right]_E. \qquad (5.07)$$

Since $g_n(x_c)=0$ and $T'(y)<0$ (see Section I.4), the second term of (5.07) is *positive* at $x=x_c$, i. e., $(\partial \ln g_n/\partial \ln x)_E<0$ at $x=x_c$. Since the first term of (5.07) is also positive, the Bessel functions in the approximate solution (5.06a) approach zero more abruptly at $x=x_c$ than do the true eigenfunctions. The discrepancy between (5.06a) and the true eigenfunctions grows with increasing x_c, since according to (1.28), the function $T'(y)\approx -(1/4)[T(0)-T(1)](2+y^{-1/2})$ approaches $-\infty$ as x goes to unity (see Section I.4). A schematic illustration of the true eigenfunction $g_0(x)$ and its Bessel-function approximation for $\sigma=0$ is given in Fig. 70 for each of three values of x_c.

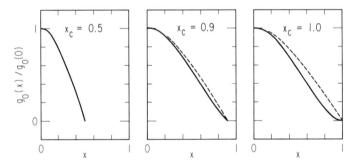

Fig. 70. Schematic representation of true eigenfunctions (solid curves) and approximate (Bessel) eigenfunctions (dashed curves) for lowest normal mode of pitch-angle diffusion with $\sigma=0$.

The use of the true pitch-angle eigenfunctions $g_n(x)$ is essential for extracting a radial-diffusion coefficient from low-altitude electron observations beyond $L\approx1.6$ (e. g., Fig. 36, Section IV.2). As shown in Section IV.2, the intensity of electrons on trajectories with "perigee" $\lesssim100\,\mathrm{km}$ increases with longitude east of the South Atlantic "anomaly", as pitch angle diffusion replenishes the pitch-angle interval $x_c<x<x_b$ (cf. Section II.7). This replenishment follows the sudden loss (by atmospheric absorption) of electrons with pitch angles $x_c<x$ as they azimuthally drift through the "anomaly" region. In effect, the loss cone seen by the

electron distribution suddenly enlarges as the particles approach the "anomaly", where $x_b = x_c$. Immediately east of the "anomaly", fewer precipitating particles are observed than immediately west of the "anomaly", since no electrons remain in the loss cone. In other words, after the excess electrons are lost in the "anomaly", the pitch angle distribution vanishes at $x = x_c$.

A reasonable method of analyzing such electron observations to obtain D_{xx} is to assume that an isotropic source term S is distributed uniformly in longitude, as in (5.05). The addition of this source term to (5.01), together with the boundary condition that $f(x,t)$ vanish for $x = x_b(\varphi)$, defines a straightforward problem of numerical analysis[48]. Any initial ($t=0$) choice of $f(x,t)$ must evolve (for $S \neq 0$) toward a periodic solution satisfying $f(x, t + 2\pi/\Omega_3) = f(x,t)$. The azimuthal coordinate φ of the particle distribution is a function of time in the sense that $\dot{\varphi} = \Omega_3$. (Recall that Ω_3 is approximately independent of x at a given energy and L value, cf. Section I.4.) The periodic solution $f(x,t) = f(x, t - 2\pi/\Omega_3)$ obtained by following the particle distribution in its azimuthal drift is equivalent to a time-independent distribution $f(x, \varphi)$, where φ is the geomagnetic longitude. The functional form of $f(x, \varphi)$ depends only upon D_{xx}/Ω_3 for a given $x_b(\varphi)$. Execution of the above-described computational program for many trial values of D_{xx}/Ω_3 should therefore yield one solution $f(x, \varphi)$ that best agrees with observations (e.g., Fig. 36, Section IV.2). The observed azimuthal variation of $\bar{f}(x, \varphi)$ thus yields a value of D_{xx}/Ω_3 and (since Ω_3 is a known function of E and L) a value for D_{xx}.

The only reported computation of this nature [77] employed $g_n(x) = (2/x_b^2)^{1/2} \cos[(2n+1)(\pi x/2x_b)]$ for the pitch-angle eigenfunctions. The results of that computation are therefore probably unreliable; sinusoidal eigenfunctions do not satisfy (5.01). The reported values for D_{xx} varied from 2×10^{-3} day^{-1} at $L=2$ to 8×10^{-3} day^{-1} at $L=4$. The corresponding electron "lifetimes" ($\approx 4/\pi^2 D_{xx}$) would amount to ~ 200 days at $L=2$ and ~ 50 days at $L=4$ for electron energy $E \approx 0.6 \pm 0.2$ MeV. Since these "lifetimes" exceed those shown in Fig. 41 (Section IV.3) by nearly an order of magnitude, the numerical values of D_{xx} on which they are based are open to question. It is difficult, of course, to rule out a possible variation of D_{xx} with x or φ that might explain the

---

[48]Due to the South American anomaly (cf. Fig. 30, III.7) which lies immediately to the west, the loss cone expands abruptly from a small aperture ($\cos^{-1} x_b$) to a larger aperture ($\cos^{-1} x_c$) in the neighborhood of the South Atlantic "anomaly". It is mathematically convenient to model the loss-cone aperture as a step function of azimuth, rather than a sinusoidal function [cf. (2.75), Section II.8]. In fact, the step function may be the more faithful representation of geophysical reality.

discrepancy[49], but a treatment based upon the true eigenfunctions of (5.01) is much needed.

Information on the possible energy dependence of D_{xx} east of the "anomaly" can be deduced by comparing the energy spectra of precipitating electrons at several longitudes $\Delta\varphi$, where $\Delta\varphi = 0$ at the "anomaly" [77]. Precipitating electrons having energies from 0.4 MeV to 2.5 MeV are found to have an exponential energy spectrum. The e-folding energy E_0 is found to increase with increasing east longitude. This observation can be understood largely in terms of the energy-dependent azimuthal-drift rates (cf. Section I.4). Thus, the pitch-angle diffusion coefficient D_{xx} apparently is not a strong function of electron energy in the range $E \approx 0.4$—2.5 MeV.

V.3 Quadrature (Spatial)

The diffusion equation can be manipulated in several ways in an attempt to extract the radial diffusion coefficient D_{LL} and/or the particle lifetime τ from the observational data. One class of methods involves a partial integration of the diffusion equation between two fixed limits in L or time. Letting $F \equiv \ln \bar{f}$ allows the radial diffusion equation [cf. (3.48), Section III.8] to be written as

$$\frac{\partial F}{\partial t} = \left[L^2 \frac{\partial}{\partial L}\left(\frac{D_{LL}}{L^2}\right)\right]\frac{\partial F}{\partial L} + D_{LL}\left[\frac{\partial^2 F}{\partial L^2} + \left(\frac{\partial F}{\partial L}\right)^2\right] - \frac{1}{\tau} \qquad (5.08)$$

for constant M and J. If the true lifetime τ is a known function of L, and if $F(L,t)$ is available from the observational data[50], then (5.08) may be interpreted as a linear first-order differential equation for D_{LL}. If the observations of F cover the interval $L_1 \lesssim L \lesssim L_2$, then the solution of (5.08) may be written

---

[49]Enhanced pitch-angle diffusion subsequent (in longitude) to "complete" replenishment of the equatorial-pitch-angle-cosine interval $x_c < x < x_b$ might escape detection by the above analytic method if it is accompanied by an enhanced source S [cf. (5.05)]. Any such enhancement of D_{xx} in longitude, however, should be correlated with magnetospheric longitude (local time) rather than geographic longitude, since atmospheric scattering of electrons is unimportant beyond $L \approx 1.5$ (cf. Fig. 41, Section IV.3, and Figs. 72—73 below).

[50]Since only derivatives of $F \equiv \ln \bar{f}$ appear in (5.08), the result is not affected by adding a constant to F. Thus, if the functional form of $\bar{f}(L,t)$ is known, the absolute normalization is not required. Only the form of the flux profile at constant M and J affects (5.08).

$$D_{LL}(L) = \exp\left[-\int_{L_3}^{L} Q_1(L')dL' \right]\left\{ D_{LL}(L_3) \right.$$
$$\left. + \int_{L_3}^{L} Q_2(L')\exp\left[\int_{L_3}^{L} Q_1(L'')dL'' \right]dL' \right\}, \tag{5.09 a}$$

where

$$Q_1(L) = \left(\frac{\partial F}{\partial L}\right)^{-1}\left[\frac{\partial^2 F}{\partial L^2} + \left(\frac{\partial F}{\partial L}\right)^2 - \frac{2}{L}\left(\frac{\partial F}{\partial L}\right)\right] \tag{5.09 b}$$

$$Q_2(L) = \left(\frac{\partial F}{\partial L}\right)^{-1}\left[\frac{\partial F}{\partial t} + \frac{1}{\tau}\right]. \tag{5.09 c}$$

The value of L_3 must lie within the range covered by the data, but is otherwise arbitrary. The value of D_{LL} at $L=L_3$ plays the role of an arbitrary integration constant. The presence of an arbitrary constant, whose value must be estimated by other means, is a persistent difficulty of analytical methods in which D_{LL} is expressed as a spatial quadrature.

The use of $F \equiv \ln \bar{f}$ rather than $\bar{f}$ itself in (5.08) is advantageous from the computational standpoint. The standard use of finite-difference techniques in evaluating an expression like (5.09) tends to introduce far less error in the derivatives of F than in the derivatives of $\bar{f}$.

The method of (5.09) is clearly inapplicable, however, if $\partial F/\partial L$ vanishes anywhere in the interval of interest $(L_1 \le L \le L_2)$. In such a case, it may be fruitful to return to (3.48), written in the form

$$\frac{\partial}{\partial L}\left[\frac{D_{LL}}{L^2}\left(\frac{\partial \bar{f}}{\partial L}\right)\right]_{M, J} = \frac{1}{L^2}\left[\frac{\bar{f}}{\tau} + \frac{\partial \bar{f}}{\partial t}\right]. \tag{5.10}$$

The full quadrature of (5.10) can be written as

$$D_{LL}(L) = L^2\left(\frac{\partial \bar{f}}{\partial L}\right)^{-1}\int_{L_3}^{L}\left[\frac{\bar{f}}{\tau} + \frac{\partial \bar{f}}{\partial t}\right]\frac{dL'}{(L')^2} + \left[\frac{D_{LL}}{L^2}\frac{\partial \bar{f}}{\partial L}\right]_{L=L_3}$$

$$= \frac{L^2}{\bar{f}}\left(\frac{\partial F}{\partial L}\right)^{-1}\int_{L_3}^{L}\left[\frac{1}{\tau} + \frac{\partial F}{\partial t}\right]\frac{\bar{f}dL'}{(L')^2} + \left[\frac{D_{LL}}{L^2}\bar{f}\left(\frac{\partial F}{\partial L}\right)\right]_{L=L_3}. \tag{5.11}$$

Here the arbitrary integration constant $D_{LL}(L_3)$ reappears. Now, however, if L_3 is chosen so that $\partial F/\partial L=0$ at $L=L_3$, then it follows from (5.11) that

$$D_{LL}(L) = \frac{L^2}{\bar{f}}\left(\frac{\partial F}{\partial L}\right)^{-1}\int_{L_3}^{L}\left[\frac{1}{\tau} + \frac{\partial F}{\partial t}\right]\frac{\bar{f}dL'}{(L')^2}. \tag{5.12}$$

Here, as in (5.09), the normalization of $\bar{f}$ does not affect the value of D_{LL} extracted from the data by integrating to an L value of interest.

When it is impossible to choose L_3 such that $(\partial F/\partial L)_{L=L_3}$ vanishes, the difficulty associated with an arbitrary integration constant can often be circumvented by postulating the analytical form of D_{LL} *a priori*. The functional form customarily postulated (subject to later verification) is a power law in L (*cf.* Sections III.2 and III.3), *i.e.*, $D_{LL}=D_n L^n$. Under this assumption, it follows from (5.08) that

$$D_n = L^{-n} \left[\frac{1}{\tau} + \frac{\partial F}{\partial t} \right] \div \left[\left(\frac{n-2}{L} \right) \frac{\partial F}{\partial L} + \frac{\partial^2 F}{\partial L^2} + \left(\frac{\partial F}{\partial L} \right)^2 \right] \qquad (5.13)$$

for some initially chosen value of n (not necessarily an integer). Since (5.13) follows from the assumption that D_{LL} is a "monomial" function of L, self-consistency can easily be checked by evaluating D_n from (5.13) for several values of L between L_1 and L_2. A moderate scatter of the resulting D_n values about some constant mean would represent a measure of uncertainty in the numerical value of D_n while confirming the postulated power law. On the other hand, a systematic variation of D_n with L would indicate that n had been chosen improperly, *i.e.*, that some other power law (or perhaps a different functional form altogether) is required in order that D_{LL} fit the data [113].

As indicated in Section IV.2, both the decay of inner-zone electron fluxes following the Starfish explosion (Fig. 32) and the decay of a monoenergetic electron enhancement observed after a magnetic storm (Fig. 35) have been adequately accounted for by atmospheric-scattering losses (Section II.2). Both sets of measurements had been made over a time interval that was relatively short compared to the calculated lifetimes. A long-term study of inner-zone electrons ($E > 0.5$ MeV), covering a three-year period beginning in September 1962, revealed considerably longer *apparent* lifetimes in the region $1.15 < L < 1.21$ than had been observed in the 50-day period immediately following the Starfish detonation. Since atmospheric scattering could not have grown abruptly less intense with time[51], these measurements suggest that additional electrons were continually being supplied to these low L shells from higher L, perhaps by radial diffusion. An empirical analysis of these data for radial-diffusion effects is facilitated by the fact that the omnidirectional flux (profile shown in Fig. 71a) decayed almost exponentially during the three-year period that began in September 1962. The *apparent* decay rate $-\partial F/\partial t$ and the decay rate $1/\tau$ "expected" on the basis of atmospheric collisions (Section II.2) are shown in Fig. 71b [114].

---

[51] However, this was a period of decreasing solar activity (see Fig. 51, Section IV.5), during which the atmosphere would have contracted toward the earth.

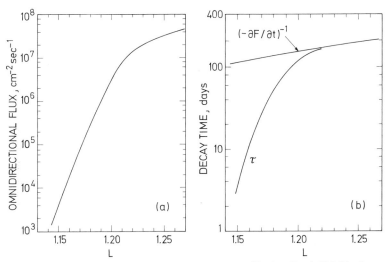

Fig. 71. (a) Inner-zone equatorial electron-flux profile for $E > 1.6$ MeV observed on 1964-45A during December 1964; (b) decay times τ and $(-\partial F/\partial t)^{-1}$ derived from atmospheric-scattering theory [42] and from a three-year compilation [123] of inner-zone electron data (1962—65; $E > 0.5$ MeV), respectively [114].

The "staircase" function shown in Fig. 72 represents a self-inconsistent determination of D_{LL} based on a hybrid analytical method [114] with features of both (5.12) and (5.13). The integral omnidirectional fluxes $I_{4\pi}$ are first converted to equatorial differential unidirectional fluxes at constant M by postulating an energy spectrum like that which results from the beta decay of fission products (cf. Section V.6) and a pitch-angle distribution compatible with the known loss-cone aperture (cf. Section II.7). The pitch-angle correction (factor converting omnidirectional flux to unidirectional flux) varies by $\sim 30\%$ between $L = 1.15$ and $L = 1.21$; it tends to reduce the slope of the flux profile. The conversion from $I_\perp$ at constant E to $\bar{f} = J_\perp/p^2$ at constant M leads to a correction that varies by $\sim 10\%$ over the interval $1.15 < L < 1.21$; this correction tends to steepen the profile. The net result is that the profile $\bar{f}$ is $\sim 20\%$ less steep than that of $I_{4\pi}$, shown in Fig. 71a.

An acceptable procedure for obtaining D_{LL} from these observational data consists of replacing $D_{LL}(L_3)$ on the right-hand side of (5.11) by $(L_3/L)^n D_{LL}(L)$. This is equivalent to assuming that $D_{LL} \propto L^n$. A rearrangement of terms then yields

$$D_{LL}(L_4) = \int_{L_3}^{L_4} \left[\frac{1}{\tau} + \frac{\partial F}{\partial t} \right] \frac{\bar{f} \, dL'}{(L')^2} \div \left[\frac{\bar{f}}{L^2} \left(\frac{L}{L_4} \right)^n \frac{\partial F}{\partial L} \right]_{L=L_3}^{L=L_4}. \tag{5.14}$$

In particular, the choice $n=0$ corresponds to a diffusion coefficient D_{LL} that is constant in the interval $L_3 \leq L \leq L_4$. The "staircase" function shown in Fig. 72 results from performing the integral in (5.14) over a sequence of consecutive, adjacent L intervals of width $L_4 - L_3 = 0.01$ for $n=0$. An alternative choice of intervals, such that $L_4 - L_3 = 0.005$ yields very similar results, as does the use of (5.13) for $n=0$.

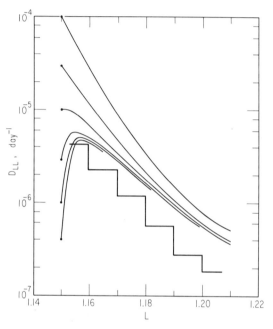

Fig. 72. Radial diffusion coefficients obtained from data in Fig. 73, assuming fission spectrum to obtain $\bar{f}(M,J,\Phi)$ at constant M and J. Staircase function [114] follows from self-inconsistent quadrature. Smooth curves are self-consistent, but require assignment of arbitrary values (filled circles) to D_{LL} at $L = 1.15$ [115].

The inconsistency of the "staircase" function as a solution for D_{LL} is that the initial assumption $(n=0)$ has led to the conclusion that D_{LL} varies inversely as $\sim L^{70}$, i.e., by a factor ~ 20 between $L=1.20$ and $L=1.15$. In fact, a treatment of the same data using (5.13) with $n = -70$ proves to be reasonably self-consistent, and leads to a value of D_{LL} that is approximately twice as large as the "staircase" function at the center of each integration interval.

The alternative procedure of assigning D_{LL} a certain arbitrary value at $L_3 = L_1 = 1.15$ yields the family of smooth curves shown in Fig. 72 [115]. Values of D_{LL} ranging from 4×10^{-7} day^{-1} to 3×10^{-5} day^{-1}

arbitrarily assigned at $L = 1.15$ thus yield remarkably similar solutions for D_{LL} beyond $L = 1.17$. The solution generated by $D_{LL}(1.15) = 1.5 \times 10^{-5}$ day^{-1} (not shown in Fig. 72) roughly approximates the above-described power law in magnitude and functional form. Of course, arbitrarily large values of $D_{LL}(L)$ could be generated for $L > L_1$ by an unreasonable choice of $D_{LL}(L_1)$.

Although the various operations on the data yield different solutions for D_{LL}, all solutions support the major conclusion that $\partial D_{LL}/\partial L < 0$ for $1.16 < L < 1.21$. This is an interesting reversal of the trend evident in observations made beyond $L = 2$ (*cf.* Section IV.6), where D_{LL} appears to vary as a large (~ 10) *positive* power of L. The reversal perhaps originates from ionospheric-current impulses [114], but a variation so extreme ($D_{LL} \propto L^{-70}$) would require very localized current distributions (spherical-harmonic number ~ 40).

Another possible origin of the reversal is atmospheric pitch-angle scattering in the presence of shell splitting caused by internal geomagnetic multipoles (Fig. 30, Section III.7). It is apparent that D_{xx} has a strongly inverse variation with L (*cf.* Fig. 71b, the curve for $1/\tau$). The analysis of Fig. 71 for such a constant-energy process must be based on (3.42), Section III.7, rather than on (3.48), since (3.48) applies to a constant-M process. The term $[(n-2)/L](\partial F/\partial L)$ in (5.13) must therefore be changed to $[(n+2)/L](\partial F/\partial L)$, for example. A somewhat larger magnitude of D_{LL} is required to account for the observations if a constant-E process is postulated instead of a constant-M process. As a rough estimate, the solutions for D_{LL} in Fig. 72 should be multiplied by a factor ~ 2 in order to accommodate a process for which inward radial diffusion does not change a particle's energy.

As noted above, the appearance of an arbitrary integration constant $D_{LL}(L_3)$ in (5.11) follows from the fact that $\partial F/\partial L$, as given by the data, fails to pass through zero in the interval $1.15 < L < 1.21$ used for analysis. The region of L over which the inner electron belt is analyzed, therefore, might profitably be extended to $L \approx 1.6$ so as to include the maximum in F that exists near $L = 1.4$ (see Fig. 73a)[52].

The apparent decay rates $-\partial F/\partial t$ are obtained from measurements made on the OV1-2 satellite, and are shown in Fig. 73b together with the calculated atmospheric-scattering lifetimes. The single point $L \approx 1.77$ in the figure is the decay lifetime of the narrow electron belt created by the Soviet nuclear detonation of 1 November 1962 (see

---

[52]These electron "distribution functions" correspond to two different values of the first invariant M [116]. Since (5.08)—(5.14) do not explicitly couple distinct values of M by differential operators, it is permissible to plot $L^3 J_\perp$ rather than $J_\perp/MB$. The calculations are unaffected by this choice.

Fig. 58, Section IV.6). The dashed curve in Fig. 73b is an "interpolation" between observed pitch-angle diffusion lifetimes of inner-belt electrons. The function $D_{LL}(L)$ obtained from (5.12) by using the lifetime data of Fig. 73b for $M = 21.4 \, \text{MeV/gauss}$ ($E = 1 \, \text{MeV}$ at $L = 1.65$) is plotted in Fig. 73b [116]. The value of D_{LL} at $L = 1.20$, as obtained from these data, is two orders of magnitude larger than the value of D_{LL} at $L = 1.20$ shown in Fig. 72. The derived magnitude of D_{LL}, however, is fairly sensitive to the numerical value assigned to τ at $L = 1.42$. According to Fig. 41 (Section IV.3), this value should have been ~ 300 days, which is much closer to the apparent lifetime $-(\partial F/\partial t)^{-1}$ than the value of τ actually used for the computation (dashed curve, Fig. 73b).

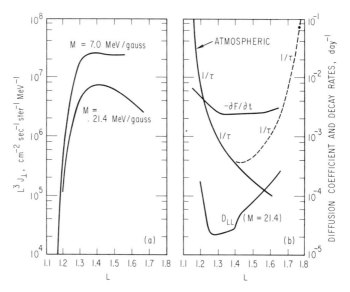

Fig. 73. (a) Profiles of electron distribution function ($\times 2m_0 M B_0$) for $J = 0$, based on OV 1-2 data; (b) decay rates $1/\tau$ expected from atmospheric scattering (solid curve) and wave-particle interaction (data point at $L \approx 1.77$); (b) arbitrary interpolation (dashed curve); (b) decay rate $(-\partial F/\partial t)$ actually observed for $E > 0.5 \, \text{MeV}$; (b) diffusion coefficient D_{LL} derived from these data for $M = 21.4 \, \text{MeV/gauss}$ [116].

The methods of this section are applicable not only to static profiles ($\partial F/\partial t = 0$) but also to time-varying profiles ($\partial F/\partial t \neq 0$). The methods can easily be modified to include the effects of a distributed source [*e.g.*, (3.57), Section III.8], as well as particle deceleration without pitch-angle diffusion. It is probably unwise, however, to attempt a purely spatial quadrature on the observed distribution of outer-zone electron fluxes. This reservation holds because outer-zone electron fluxes exhibit considerable fluctuation with time, rather than a slow evolution of the profile. Much of the observed variation is not related to radial diffusion in a simple way (*cf.* Sections IV.6 and IV.8). Even when special care is taken to select only geomagnetically quiet time intervals, the methods of this section are found to yield unreasonably large magnitudes and dubious functional forms for D_{LL} [117].

V.4 Quadrature (Temporal)

In the presence of temporal fluctuations such as those commonly observed in outer-zone electron fluxes, it is essential not only to select carefully the time interval chosen for analysis, but also to evaluate time derivatives of F from several-day averages. The interval chosen for analysis must be free of large "injection" events (*cf.* Fig. 38, Section IV.3) characterized by *in situ* particle energization, as such processes cannot easily be included in the diffusion equation. Other temporal changes in the particle fluxes, such as those due to field changes on both the adiabatic and impulsive time scales, must be averaged over time to avoid spurious contributions to $\partial F/\partial t$.

Care must be taken in obtaining the average of $\partial F/\partial t$ over several days, however, since outer-zone electron lifetimes are typically 5—10 days (*cf.* Fig. 41, Section IV.3). Thus, the averaging procedure must be sophisticated enough to accommodate the true evolution of $F(L,t)$ during the several-day time interval over which the average is taken. One procedure for performing this average, sometimes termed the *temporal quadrature* of (5.08), involves the assumption (*cf.* Section V.3) that $D_{LL} \propto L^n$. If D_n and τ are regarded as time-independent during the interval $t_1 \leq t \leq t_2$, then it follows from (5.08) that

$$D_{LL} \equiv D_n L^n = \{ F(t_2) - F(t_1) + [(t_2 - t_1)/\tau] \}$$

$$\div \int_{t_1}^{t_2} \left[\left(\frac{n-2}{L} \right) \frac{\partial F}{\partial L} + \frac{\partial^2 F}{\partial L^2} + \left(\frac{\partial F}{\partial L} \right)^2 \right] dt . \qquad (5.15)$$

The diffusion coefficient D_{LL} can thus be determined from electron data such as those shown in Fig. 74 [93]. These data have been converted to equivalent equatorial profiles of $L^3 J_\perp$ ($\equiv 2 m_0 M B_0 \bar{f}$) at constant M (*cf.* Fig. 54, Section IV.6). By choosing t_1 and t_2 appropriately, so that $F(t_2) \gtrsim F(t_1)$, it might be possible to estimate D_n for each assigned

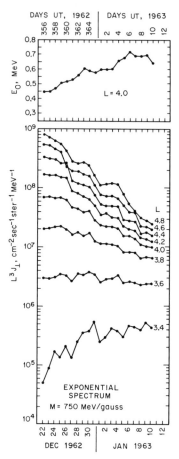

Fig. 74. Evolution of equatorial daily-median $L^3 J_\perp$ and spectral parameter following magnetic storm of 17 December 1962 [93], based on Explorer-15 electron data (cf. Figs. 53 and 54).

n. The "best" value of n would be that for which D_n is most nearly independent of L (cf. Section V.3).

Unfortunately, the application of (5.15) to the full twenty-day interval of data shown in Fig. 74 does not allow a precise determination of D_{LL}. This is because the quantity $F(t_2) - F(t_1)$ is negative at each L value shown (as is usual for such a long time interval) and represents a good approximation for $(t_1 - t_2)/\tau$. The numerator of (5.15) is therefore approximately zero, and so is very sensitive to the somewhat arbitrary choice of lifetime $\tau(L)$. This difficulty arises quite frequently in practice, since (as noted above) $t_2 - t_1$ must be chosen sufficiently long to average out the adiabatic fluctuations[53] in $\partial F/\partial t$.

An alternative to the above procedure is to solve (5.15) for the pitch-angle-diffusion lifetime $\tau(L)$, in terms of the radial diffusion coefficient $D_{LL} = D_n L^n$. In this case, time-independent values of n and D_n are chosen somewhat arbitrarily in order to obtain

$$\frac{1}{\tau} = \frac{D_n L^n}{t_2 - t_1} \int_{t_1}^{t_2} \left[\left(\frac{n-2}{L} \right) \frac{\partial F}{\partial L} + \frac{\partial^2 F}{\partial L^2} + \left(\frac{\partial F}{\partial L} \right)^2 \right] dt - \frac{F(t_2) - F(t_1)}{t_2 - t_1}. \quad (5.16)$$

For use in (5.16) the observational data shown in Fig. 74 can be manipulated to yield numerical derivatives given by the algebraic expressions $F'(L;t) = (5/2)[F(L+0.2;t) - F(L-0.2;t)]$, $F''(L;t) = 25[F(L+0.2;t) - 2F(L;t) + F(L-0.2;t)]$ and $\dot{F}(L,t) = (1/2)[F(L;t+1) - F(L;t-1)]$,

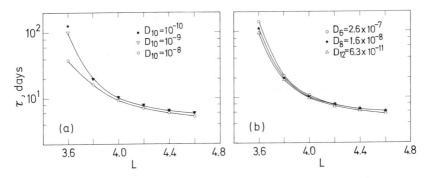

Fig. 75. Electron lifetimes ($M = 750$ MeV/gauss) obtained from data of Fig. 74 by using (5.16) for selected values of D_n, day^{-1}.

with time t measured in days. Numerical integration of (5.16) by Simpson's rule then yields the lifetimes shown in Fig. 75a.

---

[53]The fluctuations superimposed on the mean evolution in Fig. 74 are found to be well correlated with the ring-current index D_{st} (cf. Section I.5). However, attempts to suppress these apparently adiabatic fluctuations by means of a ring-current model (cf. Fig. 9, Section I.5) and available spatial and spectral information (cf. Fig. 43, Section IV.3) did not succeed for the data of Fig. 74. Perhaps the modeling procedures commonly used for protons (Fig. 43) are insufficiently accurate to subtract the adiabatic fluctuations of outer-zone electrons with confidence.

The functions $\tau(L)$ shown in Fig. 75a are very similar, although the inserted values of D_{10} vary over two orders of magnitude. The choice of $n = 10$ follows a convention based on "theoretical" considerations (cf. Sections III.2, III.3, and III.8). Since all choices from 10^{-10} day^{-1} to 10^{-8} day^{-1} for the magnitude of D_{10} are found to yield equally reasonable lifetime functions $\tau(L)$, the "correct" magnitude for D_{10} must be obtained by invoking some further empirical consideration.

Values of $n \neq 10$ yield similar $\tau(L)$ functions to those shown in Fig. 75a if D_n is chosen appropriately. Since the observations cover a range centered at $L \approx 4$, a logical comparison among different values of n would require that $D_n 4^n$ be held fixed. Thus, the $\tau(L)$ functions plotted in Fig. 75b for $n = 6$, 8, and 12 (with $D_n = D_{10} 4^{10-n}$) are virtually indistinguishable.

If the data of Fig. 74 are analyzed in blocks of five days instead of twenty, (i. e., $t_2 - t_1 = 5$ days) the application of (5.16) to these separate intervals is found to yield lifetime functions $\tau(L)$ that duplicate Fig. 75 within a factor of two. Thus, while temporal quadrature yields self-consistent lifetimes against pitch-angle diffusion, the extraction of a radial-diffusion coefficient D_{LL} from data such as shown in Fig. 74 apparently requires another (more sophisticated) analytical technique.

V.5 Variational Method

A major disadvantage of quadrature (either spatial or temporal, cf. Sections V.3 and V.4) in the extraction of transport coefficients from time-varying electron data is that either D_{LL} or D_{xx} must be given a priori in order to obtain the other. An empirical technique termed the "variational method" circumvents this difficulty and thus enables both the radial-diffusion coefficient D_{LL} and the particle-lifetime function $\tau(L)$ to be extracted simultaneously from the data with minimum reliance on ad hoc assumptions about the L dependence of τ.

The variational technique involves the usual tacit assumptions that both τ and D_{LL} are time-independent and that D_{LL} can be represented in the form $D_{LL} = D_n L^n$. Then the temporal evolution of $F(L, t)$, as given by (5.08), can be attributed to a combination of diffusion across L and pitch-angle scattering into the loss cone. To the extent that radial diffusion can be accounted for by properly choosing the magnitude and functional form of D_{LL}, the remaining temporal decay of F via pitch-angle scattering should be linear, corresponding to an exponential decay of $\bar{f}(L, t)$. The idea of the variational method [93] is to formulate a quantitative measure of the extent to which a given D_n "fails" to account for the nonlinear temporal component of the evolution of $F(L,t)$.

This formulation is facilitated by introducing the decay-rate function [*cf.* (5.16), Section V.4]

$$\lambda_n(L, t) \equiv D_n L^n \left[\left(\frac{n-2}{L} \right) \frac{\partial F}{\partial L} + \frac{\partial^2 F}{\partial L^2} + \left(\frac{\partial F}{\partial L} \right)^2 \right] - \frac{\partial F}{\partial t}. \qquad (5.17)$$

This function reduces to a constant in time [*viz.*, $1/\tau(L)$] only if $F(L,t)$, as given by the data, exactly satisfies (5.08). The "correct" value of n is that which enables D_n to be chosen so that $\lambda_n(L,t)$ is constant in time, and the "correct" value of D_n is that which makes $(\partial \lambda_n/\partial t)_L$ vanish.

In practice, of course, there will be uncertainties in the data, and it may be impossible to suppress adiabatic fluctuations satisfactorily (*cf.* Section V.4). These and other difficulties prevent $\lambda_n(L,t)$, as given by the data, from being *exactly* constant in time under any conditions. It is possible, however, to ask (for any given n) that D_n be chosen so that $\lambda_n(L,t)$ *deviates minimally* from a constant. The deviation of $\lambda_n(L,t)$ from a constant in time can be expressed quantitatively by introducing a function

$$G_n(D_n) \equiv \int\limits_{L_1}^{L_2} g(L) \int\limits_{t_1}^{t_2} \left[\lambda_n^2 - \langle \lambda_n \rangle^2 \right] dt \, dL, \qquad (5.18)$$

where $\langle \lambda_n \rangle$ is the temporal mean value of $\lambda_n(L,t)$ and $g(L)$ is a positive-definite weighting function[54]. The function $G_n(D_n)$ is thus a quantitative measure of the "failure" of a specific numerical value of D_n to account for the time variation of F attributable to radial diffusion. The function $G_n(D_n)$ is minimized with respect to its argument (D_n) by requiring that

$$\frac{\partial G_n}{\partial D_n} = 2 \int\limits_{L_1}^{L_2} g(L) \int\limits_{t_1}^{t_2} \left[\lambda_n \frac{\partial \lambda_n}{\partial D_n} - \langle \lambda_n \rangle \frac{\partial \langle \lambda_n \rangle}{\partial D_n} \right] dt \, dL = 0. \qquad (5.19)$$

This linear algebraic equation for D_n yields a numerical value of D_n that is uniquely determined by the data for a given weighting function $g(L)$. The optimal (G_n-minimizing) value of D_n is given by

---

[54]The purpose of $g(L)$ is to distribute responsibility for the ultimate determination of D_n equitably among the various L values (see below).

$$D_n = \int_{L_1}^{L_2} L^n g(L) \int_{t_1}^{t_2} \left\{ \frac{\partial F}{\partial t} \left[\left(\frac{n-2}{L}\right) \frac{\partial F}{\partial L} + \frac{\partial^2 F}{\partial L^2} + \left(\frac{\partial F}{\partial L}\right)^2 \right] \right.$$

$$\left. - \left[\frac{F(t_2) - F(t_1)}{t_2 - t_1} \right] \left\langle \left(\frac{n-2}{L}\right) \frac{\partial F}{\partial L} + \frac{\partial^2 F}{\partial L^2} + \left(\frac{\partial F}{\partial L}\right)^2 \right\rangle \right\} dt \, dL$$

$$\div \int_{L_1}^{L_2} L^{2n} g(L) \int_{t_1}^{t_2} \left\{ \left[\left(\frac{n-2}{L}\right) \frac{\partial F}{\partial L} + \frac{\partial^2 F}{\partial L^2} + \left(\frac{\partial F}{\partial L}\right)^2 \right]^2 \right. \tag{5.20}$$

$$\left. - \left\langle \left(\frac{n-2}{L}\right) \frac{\partial F}{\partial L} + \frac{\partial^2 F}{\partial L^2} + \left(\frac{\partial F}{\partial L}\right)^2 \right\rangle^2 \right\} dt \, dL,$$

where the angle brackets denote a time average over the interval $t_1 \leq t \leq t_2$. The calculation of an optimal D_n by (5.20) can be carried out for each of many values of n. Insertion of the optimal D_n in (5.16) allows a determination of the decay constant $1/\tau(L) \equiv \langle \lambda_n(L,t) \rangle$. The best-fitting functional form of $D_{LL} = D_n L^n$ can perhaps be identified by searching for an absolute minimum in $G_n(D_n)$ among the various values of n. If some n is clearly identified as the optimum, the corresponding $\langle \lambda_n(L,t) \rangle^{-1}$ is simultaneously established (although tentatively, cf. Section V.6) as the optimum lifetime function $\tau(L)$. In practice [93], many values of n yield almost identical minima in $G_n(D_n)$. The optimal value of D_n is thus identified for each n, but the best value of n remains unidentified (cf. Section V.4).

The data of Fig. 74 (Section V.4) were extracted from flux profiles such as those in Fig. 53 (Section IV.6) by assuming an exponential energy spectrum at each L value. Very similar data representing $L^3 J_\perp = 2m_0 M B_0 \bar{f}$ are obtained by postulating a power-law spectrum

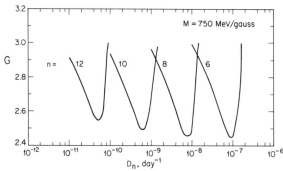

Fig. 76. Functions $G_n(D_n)$ obtained for $g(L) \equiv 1$, from data similar to those of Fig. 74 but using a power-law spectrum (cf. Fig. 54) for energy interpolation [93].

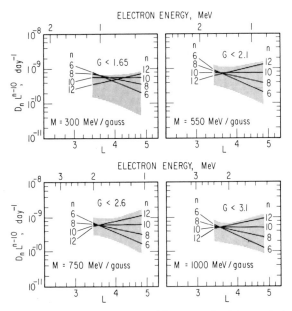

Fig. 77. Optimal values (solid lines) of $L^{-10} D_{LL}$ obtained by variational method, with $g(L) \equiv 1$, from data of Fig. 74. Shaded area represents range of values acceptable [93] in the context of Fig. 76.

for interpolation between $E = 0.5$ MeV and $E = 1.9$ MeV (*cf.* Fig. 53 [93]). The functions $G_n(D_n)$ constructed [using (5.18)] from these power-law data are plotted in Fig. 76 for several values of n with $g(L) \equiv 1$. It is evident from Fig. 76 that the optimal (G_n-minimizing) value of D_n is easily identified for any given n. On the other hand, the several values of n yield virtually identical minimum values for the functions $G_n(D_n)$. The failure of the variational method to yield a unique optimum value of n in this case is perhaps a consequence of the narrowness of the interval in L available for analysis ($L_2 - L_1 = 1.4$; $L_2/L_1 = 1.4$).

The optimal values of $D_n L^{n-10}$ obtained from the analyses [*cf.* (5.20)] for several values of n and four values of M are shown in Fig. 77 by solid lines. The shaded area contains all values of D_{LL}/L^{10} such that $6 \leq n \leq 12$ and $G_n(D_n)$ is less than the stated limit (*e. g.*, $G_n < 2.6$ at $M = 750$ MeV/gauss). The several values of n thus yield a fairly consistent value of D_{LL} at $L \approx 3.6$ (*cf.* Section V.4). The lifetime functions $\tau(L)$ are obtained for each M by inserting in (5.16) the optimal values of D_n, as obtained from (5.20). The results are shown in Fig. 78 [93].

The variational method is a good technique for extracting numerical values of the transport coefficients D_{LL} and τ from observational data

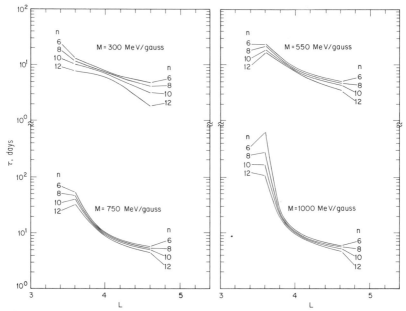

Fig. 78. Electron lifetimes obtained from (5.16) by inserting optimal values of D_n indicated in Fig. 77 [93].

consisting of time-varying flux profiles[55]. When used in conjunction with one of the verifying methods described below (Section V.6), the variational method constitutes a systematic and expedient means of analyzing the observational data for simultaneous radial and pitch-angle diffusion.

Certain refinements of the variational method merit further attention. Since $\langle \lambda_n(L,t) \rangle$ is found to vary by an order of magnitude between $L=3.6$ and $L=4.6$ (cf. Figs. 75 and 78), use of the weighting function $g(L) \equiv 1$ in (5.18) tends to leave $G_n(D_n)$ relatively insensitive to observations made at $L \lesssim 4$. Thus, in identifying the optimal value of D_n, undue weight is perhaps assigned to the region $L \gtrsim 4$. An alternative weighting function of the form $g(L) \equiv 65/L^3$, normalized to a unit mean value in the interval $3.4 < L < 4.8$, partially redresses the imbalance and introduces no significant change in Figs. 76—78 [93]. However, a weighting function that more fully compensates for the L dependence of τ might be more suitable.

The variational method is potentially sensitive to genuine temporal variations of the transport coefficients D_{LL} and τ. There is no provision

---

[55] Note that (5.20) is indeterminate for any time interval $t_1 \leq t \leq t_2$ in which $\partial F/\partial t \equiv 0$.

in (5.18)—(5.20) for recognizing such variations. Moreover, if the data are rather "noisy" as in Fig. 74, containing fluctuations unrelated to radial and pitch-angle diffusion, the equations may seek to minimize $G_n(D_n)$ not by selecting the best D_{LL} and $\langle \lambda_n(L,t) \rangle$, but by selecting unrealistically small values of $\langle \lambda_n(L,t) \rangle$. Such a selection would also tend to underestimate the magnitude of D_{LL} appropriate for (5.08). For these and other reasons, the magnitudes of D_{LL} and $\tau(L)$ that follow from (5.20) and (5.16), respectively, should be regarded as tentative choices. It remains to verify that, when these numerical values are inserted into (5.08), the actual evolution of $F(L,t)$ is correctly predicted by integrating this diffusion equation with respect to time (cf. Section V.6).

V.6 Temporal Integration

As described in the previous section, estimates of the radial and pitch-angle diffusion coefficients can be extracted from observed time variations in the electron fluxes. Verification of any proposed set of numerical values for D_{LL} and $\tau(L)$, whether obtained from the variational method or otherwise, requires that (5.08), the diffusion equation, be integrated with respect to time, using appropriately selected boundary conditions on $F(L,t)$. Initial conditions are determined, as far as possible, from the observational data. Given the initial and boundary conditions, the diffusion equation can then be integrated with respect to time. The result of this integration should be compared to the observed evolution of $F(L,t)$. Source terms are generally omitted for outer-zone electrons, since the observed variations in flux are presumed to occur after the source that produced the initial flux enhancement is turned off.

The inward-moving "edge" of the flux profile shown in Fig. 52a, Section IV.6 (from the same time period as the data in Fig. 74 treated above; see Sections V.4 and V.5), can be studied further by integrating (5.08), the diffusion equation, for $F(L,t)$. Since the observations consist of flux measurements for one energy threshold only, it is necessary to introduce assumptions as to the shape of the electron energy spectrum at $t=0$ (20 December). In order to obtain $F(L,0)$ at constant first invariant M, the initial energy spectrum is assumed to be exponential (with an e-folding energy of 600 keV) at $L=4$ (cf. Fig. 74, Section V.4), and consistent with (4.01), Section IV.5, at other L values.

The boundary conditions used in one analysis of these data are specified by extrapolating the $t=0$ distribution function smoothly to zero at $L=1$ and $L=8$. These boundary conditions are maintained throughout the computation. The decay time for electrons is taken as a constant, equal to 20 days, independent of time, L, and M. The M-independent radial diffusion coefficient is assumed to be independent of time also (as in the variational approach) and to have an L^n power-law dependence. The integration of (5.08), using the above-specified spectrum, boundary conditions, and lifetimes, is performed by standard finite-difference techniques. The results for $F(L,t)$ are then converted back to integral omnidirectional fluxes ($E > 1.6$ MeV) for comparison with the observational data.

Results of the computation for each of two different values of D_{LL} are compared to the observational data in Fig. 79 [70]. It is difficult

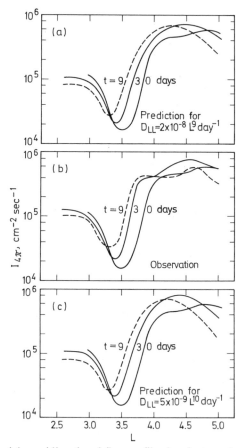

Fig. 79. Equatorial omnidirectional-flux profiles for electrons having $E > 1.6$ MeV: (a, c) predictions generated by numerical integration from $t=0$ (20 December 1962) for $\tau = 20$ days and D_{LL} as indicated, using fixed boundary conditions ($\bar{f} = 0$) at $L=1$ and $L=8$ [70]; (b) observations based on Explorer-14 data, shown also in Fig. 52a [111].

to choose the "better" D_{LL} from this comparison, but the slopes of the "leading edges" of the observed and calculated flux profiles possibly agree better for $n=10$ than for $n=9$. Given the *ad hoc* assumptions, the agreement is found to be much poorer for other values of D_n and for other integer values of n. The two "best" diffusion coefficients D_{LL} agree in magnitude at $L \approx 4$ (*cf.* Section V.4).

The validity of the results obtained by this method can presumably be tested by studying the sensitivity of these results to changes in the assumed initial conditions, boundary conditions, and lifetimes. Reasonable changes in the boundary conditions at $L>5$ (but not at $L<2.6$) are found to produce significant modifications in the results, but only in the region $L>4$ (not in the "leading-edge" region). The use of an initial *e*-folding energy of 400 keV or 800 keV, rather than 600 keV, at $L=4$ is found to cause significant changes in the calculated fluxes at L values beyond the "leading edge" of the profile, but this does not alter the overall time evolution. The optimum radial diffusion coefficient, as obtained by this integration scheme, is $D_{LL} \approx 5 \times 10^{-9} L^{10}$ day^{-1}, a result that is largely insensitive to minor variations in the multitude of qualifying assumptions indicated.

A similar analysis can be made using measurements of the artificially produced "spike" of electron flux shown in Fig. 58 (Section IV.6). In this case, the initial flux distribution is taken as an approximate Gaussian in L, with boundary conditions selected such that the fluxes are held equal to zero at $L=1.6$ and $L=1.9$. In order to obtain a constant-M distribution function $\bar{f} \equiv \exp F$, the energy spectrum is assumed to have the form [118]

$$J_{\perp}(E) \propto (v/c) \exp[-0.2938(\gamma-1) - 0.0144(\gamma-1)^2] \qquad (5.21)$$

at $L=1.765$. This is the spectrum of energies that results from the beta decay of nuclear-fission products in equilibrium. The pitch-angle-diffusion lifetimes $\tau(L)$ are assumed to be given by linear interpolation between $\tau(1.5)=470$ days and $\tau(2.1)=20$ days (*cf.* Fig. 41, Section IV.3) for all values of M. When (5.08) is solved for the evolution of $F(L,t)$ using each of several trial values of D_{LL}, it is found that the best agreement with observation corresponds to $D_{LL} \approx 6 \times 10^{-6}$ day^{-1} at $L=1.76$ [70]. This value is identical with the magnitude $\sim 6 \times 10^{-6}$ day^{-1} obtained from the analysis assuming conservation of particle energy rather than M and J (see Section IV.6).

There exists an alternative to the arbitrary imposition of boundary conditions outside the interval of L covered by the data. It is possible instead to obtain a realistic set of time-dependent boundary conditions directly from the data. Imposed at L_1 and L_2 ($=3.4$ and 4.8, respectively, in Fig. 74, Section V.4), these boundary conditions are suitable for testing the validity of a tentatively established set of transport coefficients D_{LL} and $\tau(L)$ via temporal integration of (5.08).

The initial conditions for this temporal integration of (5.08) are given by the observational values of $F(L,t_1)$, where t_1 corresponds to 22 December. Two apparently incompatible sets of transport coefficients have been identified above for electrons having $M=750$ MeV/gauss following the December 1962 magnetic storm. The variational method (Section V.5) yields $D_{LL}=6\times10^{-10}L^{10}$ day^{-1} and $\tau=\tau(L)$, as given in Fig. 78, for the twenty-day interval beginning 22 December. The method of temporal integration with fixed boundary conditions outside the data interval (Section V.6) yields $D_{LL}=5\times10^{-9}$ L^{10} day^{-1} for an assumed L-independent lifetime of 20 days over the ten-day interval beginning 20 December.

The results of temporal integration from $t=t_1$ (22 December) for the evolution of $F(L,t)$, using the observed values of $F(L_1,t)$ and $F(L_2,t)$ as time-dependent boundary conditions, are shown in Fig. 80 [119]. The predictions based on the smaller D_{LL} (obtained by the variational method) are clearly in better agreement with the observational data $(3.6\leq L\leq4.6)$ than the predictions based on $D_{LL}=5\times10^{-9}$ L^{10} day^{-1}, when the twenty-day interval is viewed as a whole. Only during the first few days of the integration interval (i.e., prior to Day 360) is there a hint that the value $D_{LL}=6\times10^{-10}$ L^{10} day^{-1} might be inadequate.

According to Fig. 79, the larger value of $D_{LL}=5\times10^{-9}L^{10}$ day^{-1} should have applied only to the ten-day interval beginning 20 December, rather than the twenty-day interval beginning 22 December. However, the lifetime function $\tau(L)$, as given in Fig. 75 (Section V.4), is found to vary by at least a factor of seven between $L=3.6$ and $L=4.6$ for any reasonable choice of D_{LL}. Thus, it is appropriate to test $D_{LL}=5\times10^{-9}$ L^{10} day^{-1} in conjunction with the lifetime function $\tau(L)$, as given in Fig. 75 or 78. For $\tau(L)$ given by Fig. 78 and $D_{10}=5\times10^{-9}$ day^{-1}, the temporal integration of (5.08) with time-dependent boundary conditions at $L_1=3.4$ and $L_2=4.8$ from $t=0$ (20 December) is found to produce good agreement with the observations $(3.6\leq L\leq4.6)$ until about 25 December (see Fig. 81). During this six-day interval, the smaller value of $D_{10}=6\times10^{-10}$ day^{-1} is clearly inadequate to account for the continuing growth of $F(L,t)$ at the lower L values $(L\lesssim4.2)$. The discrepancy beyond 25 December can be eliminated by reverting to the smaller value of D_{10} (cf. Fig. 80).

These results clearly demonstrate that a time dependence of D_{LL} was associated with the large magnetic storm of 17—18 December 1962. The choice of 22 December as t_1 apparently eliminates most of the storm-time effects that would invalidate the variational method

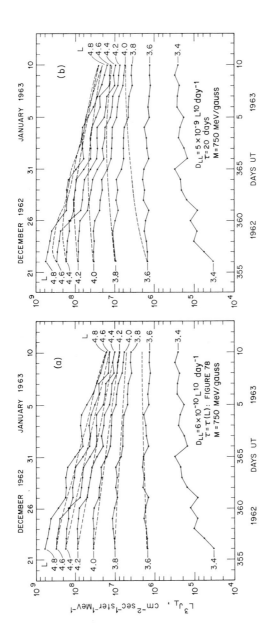

Fig. 80. Evolution of $L^3 J_\perp$ for outer-zone electrons, beginning with 22 December 1962. Observational data points joined by solid line segments are taken from Fig. 74. Dashed curves are predictions generated by numerical integration with time-dependent boundary conditions imposed by the data at $L=3.4$ and $L=4.8$ [119].

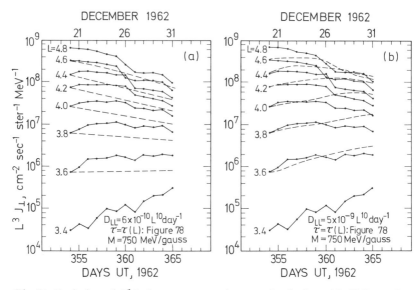

Fig. 81. Evolution of $L^3 J_\perp$ for outer-zone electrons, beginning with 20 December 1962. Observational data points are joined by solid line segments. Dashed curves are predictions generated by numerical integration with time-dependent boundary conditions imposed by the data at $L = 3.4$ and $L = 4.8$ (cf. Fig. 80).

in its present form (Section V.5). Application of the variational method to the time interval 20—24 December would perhaps yield the larger diffusion coefficient D_{LL} evidently required for that interval, if by some means the data were smoothed to suppress adiabatic and other temporal fluctuations not associated with the radial and pitch-angle diffusion processes. It is evidently impossible, however, to characterize the radial diffusion coefficient as a universal constant that can be applied uncritically to time periods during and following all magnetic storms.

A set of proton measurements, made with a scintillation counter on the elliptical-orbit satellite Explorer 26, revealed time variations in the outer-zone proton fluxes following the magnetic storm of 18 April 1965. After removal of the adiabatic variations (due to the storm-time ring current, cf. Section I.5) from the observations, non-adiabatic changes in flux were found to have occurred during the storm. Thereafter the fluxes slowly recovered non-adiabatically to their pre-storm levels. These non-adiabatic post-storm observations can be attributed to radial diffusion and atmospheric collisions.

Examples of adiabatically-corrected $L^3 \bar{J}_\perp / M$ profiles for equatorially-mirroring protons at several values of M were shown in Fig. 43 (Section IV.3). The data from several energy channels were used to construct these constant-M distribution functions, whose temporal evolution can be used to estimate a numerical value for the radial diffusion coefficient. In this case the diffusion equation has two separate loss terms. The first represents Coulomb energy loss [see (2.04) and (2.06), Section II.2; also (3.57), Section III.8]. The second loss term is equal to $-\bar{f}/\tau_q$, where τ_q is the mean proton lifetime against charge exchange [see (2.09), Section II.2]. Pitch-angle scattering of these outer-zone protons by plasma waves is apparently negligible except at $M=45\,\text{MeV}/$ gauss.

The Fokker-Planck equation governing outer-zone protons having $M \gtrsim 100\,\text{MeV/gauss}$ and $J=0$ is thus of the form

$$\frac{\partial \bar{f}}{\partial t} = L^2 \frac{\partial}{\partial L}\left[\frac{D_{LL}}{L^2}\frac{\partial \bar{f}}{\partial L}\right]_M + \frac{(4\pi q^4/m_e)}{(2\,M\,B_0^3/L^9\,m_p)^{1/2}}\left[\frac{\partial(C\bar{f})}{\partial M}\right]_L - \frac{\bar{f}}{\tau_q}, \qquad (5.22)$$

where C is given by (3.57 b). Time-dependent boundary conditions for the solution of (5.22) are imposed by the observational data at $L_1=2.1$ and $L_2=5.6$. Preliminary results suggest that an M-independent radial-diffusion coefficient $D_{LL} \sim 1\times 10^{-9}\,L^{10}\,\text{day}^{-1}$ adequately accounts for the observed temporal evolution of $\bar{f}(M,L;t)$ at $J=0$ for $L_1<L<L_2$ [82].

V.7 Spatial Integration

When the observational data consist of time-independent flux profiles $\bar{J}_\perp(E,L)$, it is necessary to obtain the relevant time-independent solution of the Fokker-Planck equation. If the transport coefficients and boundary conditions are time-independent, then the solution $\bar{f}(M,J,L;t)$ for (5.22) and similar equations will ultimately approach the steady-state solution $\bar{f}(M,J,L;\infty)$ after a sufficiently long integration time. In many situations, however, it is computationally more practical to dispense with temporal integration altogether by setting $\partial\bar{f}/\partial t=0$ at the outset. The result is a partial differential equation in the variables L and M (perhaps also J). One problem often treated in this manner is that of the inner proton belt, as described by (3.57) [see Section III.8].

Substantial theoretical effort has been expended on identifying the possible sources of the high-energy ($E \gtrsim 20\,\text{MeV}$) proton radiation observed in the inner zone at $L \lesssim 2$. Much of this theoretical work has focused on attempts to vindicate the decay of cosmic-ray-produced albedo neutrons (CRAND) as the predominant source (see Section III.8). When radial diffusion is neglected, it is found that the CRAND-source hypothesis cannot successfully account for the observed absolute intensities, nor the spatial and spectral distributions of inner-zone protons. However, a reasonable fit to the observed high-energy inner-zone proton distributions can be obtained when radial diffusion and the geomagnetic secular variation (see Section II.2) are allowed to operate on protons injected by the CRAND source.

The most extensive inner-zone proton data assembled to date were obtained by a set of shielded semiconductor detectors flown on the United States Air Force satellite OV 3-4 as an investigation for biological purposes. These integral proton-flux data, measured above five energy thresholds (15 MeV, 30 MeV, 55 MeV, 105 MeV, and 170 MeV), can be converted to equivalent equatorial profiles of $\bar{J}_\perp/MB \equiv 2m_0\bar{f}$ for selected values of the first invariant M. The results are plotted as the data points in Fig. 82.

If the geomagnetic secular variation is tentatively neglected, the time derivative $\partial\bar{f}/\partial t$ appearing in (3.57) can be set equal to zero in the search for a steady-state distribution $\bar{f}(M,L)$. The source term S is considered to be given by (3.56), and the loss term represents proton

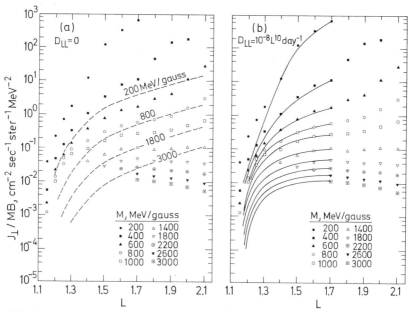

Fig. 82. Inner-zone proton distribution function $(\times 2m_0)$ for $J=0$ and selected values of M, based on OV 3-4 data and numerical integration. Dashed curves (a) are steady-state solutions of (3.57) for $D_{LL}=0$ and S given by (3.56). Solid curves (b) are steady-state solutions of (3.57) for $D_{LL}=10^{-8}\ L^{10}$ day^{-1}, with S given by (3.56) and boundary conditions imposed by the data at $L=1.1$ (where $\bar{f}=0$) and $L=1.7$ [38].

energy loss by collisions with free and bound atmospheric electrons (see Section II.2). Charge exchange is a negligible process at the proton energies of interest (*cf.* Fig. 15, Section II.2).

If radial diffusion is neglected (by taking $D_{LL}=0$), then (3.57) becomes an ordinary first-order differential equation for $\bar{f}(M)$ at each L value. A unique solution is obtained by requiring that $C\bar{f}$ vanish in the limit $M=\infty$. This solution, indicated by the dashed curves in Fig. 82a, bears little resemblance to the observational data [38].

In the presence of a nonvanishing D_{LL}, it becomes necessary to specify boundary conditions in L as well as in M. Since the purpose is to verify the adequacy of the CRAND source, the lower boundary condition should be that $f(M,L)$ vanish at some $L=L_1\approx 1.10$. This lower boundary condition identifies the dense atmosphere as a sink for the inner-belt protons, and yet does not conflict with the observation that $\bar{f}(M,L)\neq 0$ at $L=1.15$. The upper boundary condition is imposed by the observational data at $L_2=1.70$, beyond which temporal variations of $\bar{f}(M,L)$ are known to occur. For computational convenience it is assumed that $\bar{f}(M,L)=0$ at $M=4\,\text{GeV/gauss}$ (rather than $M=\infty$) throughout the interval $L_1\leq L\leq L_2$. The solution thus obtained by choosing $D_{LL}=1\times 10^{-8}\,L^{10}\,\text{day}^{-1}$ is indicated by the solid curves in Fig. 82b [38]. A vast improvement in the agreement between theory and observation is thus obtained by allowing the CRAND source to be complemented by protons diffusing inward from the outer zone.

A further improvement is expected to follow [39] from inclusion of the geomagnetic secular variation (see Section II.2). A correct treatment will require the use of (3.57) in its time-dependent form, with the Coulomb energy-loss rate expressed as a function of M, Φ, and t. It may be difficult to model the time dependence of $D_{\Phi\Phi}$ and the boundary conditions on $\bar{f}(M,L;t)$ over a history that extends at least back to Biblical times, but such an extrapolation seems necessary in order to account fully for protons now present in the inner belt. Even with the secular effect omitted (as in Fig. 82b), however, reasonable agreement between theory and observation has been obtained by allowing the generally accepted CRAND proton source [72] to operate in the presence of radial diffusion[56].

---

[56]Recent measurements suggest that the actual neutron flux exceeds the previously accepted value [72] by a factor ~ 50 at each energy above $\sim 50\,\text{MeV}$ [87].

VI. Summary

The dynamical processes that govern the earth's radiation belts have a phenomenological representation in the form of a modified Fokker-Planck equation. The operand of this partial differential equation is the drift-averaged phase-space distribution function $\bar{f}=\bar{J}_\perp/p^2$, where $J_\perp$ is the differential unidirectional particle flux at pitch angle $\pi/2$ and p is the scalar momentum. The transport coefficients that enter the modified Fokker-Planck equation correspond to the violation of one or more of the three adiabatic invariants of charged-particle motion. Adiabatic invariants of radiation-belt particles can be violated either by collisions with atmospheric constituents or by interactions with magnetospheric waves and other disturbances. For processes not involving collisions of radiation-belt particles with the atmosphere, the Fokker-Planck equation typically reduces to a diffusion equation. The basic objective of radiation-belt physics is to account for the observational data on energetic particles within the context of the Fokker-Planck equation by inserting realistic transport coefficients for the operative dynamical processes.

Non-Diffusive Phenomena. The principal effect of the atmosphere on radiation-belt protons and heavier ions is to cause charge exchange and deceleration through Coulomb collisions. Charge-exchange collisions of protons can be represented by a simple loss term of the form $-\bar{f}/\tau_q$. However, for multiply-charged ions (mainly helium in the case of the earth's magnetosphere), charge exchange serves to couple the transport equations governing the distribution functions of variously charged ions derived from identical nuclei. Charge exchange and the deceleration of ions by Coulomb collisions are accompanied by very little pitch-angle diffusion or energy diffusion (range straggling).

Pitch-Angle Diffusion. The atmospheric scattering of radiation-belt electrons involves diffusion in both pitch angle and energy, as well as deceleration. The Coulomb collisions that are important in the context of deceleration and range straggling involve the individual electrons that surround the nucleus. Pitch-angle diffusion involves the nuclear charge, as shielded by the electron cloud. Only at $L \lesssim 1.3$ do atmospheric collisions importantly affect electrons that mirror near the geomagnetic equator. Atmospheric scattering of electrons at $L \gtrsim 1.3$ is important only in that it defines a loss cone in velocity space for electrons acted upon by other dynamical processes.

Beyond $L \approx 1.3$, electron pitch angles diffuse into the atmospheric loss cone primarily by virtue of interactions between the electrons and magnetospheric plasma waves. Except for electrons that mirror very near the magnetic equator, the interaction that produces pitch-angle diffusion is probably a Doppler-shifted cyclotron resonance with electromagnetic cyclotron waves (whistler mode). The pitch-angle diffusion of electrons mirroring near the equator can be produced by cyclotron resonance with electrostatic cyclotron waves (Bernstein modes), by bounce resonance with MHD waves, or by Landau resonance ($\omega = k_{\parallel} v_{\parallel}$) with whistler-mode waves propagating at oblique angles to the geomagnetic field.

The empirical result of wave-particle interactions is a pitch-angle-diffusion lifetime that varies strongly with L, but only moderately with energy, for radiation-belt electrons. The lifetime at fixed energy ($E \sim 0.5 \, \text{MeV}$) typically decreases by a factor ~ 60 (from ~ 300 days to ~ 5 days) between $L = 1.5$ and $L = 5$ (cf. Figs. 39—41, Section IV.3). Electrons having $E \sim 1 \, \text{MeV}$ are more long-lived ($\tau \sim 10$ days at $L \sim 5$). It is plausible that the "slot" between the inner and outer electron belts (at $L \sim 3$) may correspond to a region of anomalously intense pitch-angle diffusion[57]. Quantitative evaluation of pitch-angle diffusion for electrons inside the plasmasphere, resulting from resonant interactions with obliquely-propagating whistler-mode waves, has provided reasonable agreement with observed electron lifetimes [121].

Proton pitch-angle diffusion, presumably caused by electromagnetic ion-cyclotron waves, is believed to play an important role in the dynamics of radiation-belt protons having $E \lesssim 400 \, \text{keV}$. Isolated instances of pitch-angle diffusion have been detected also for protons with $E \sim 5$—$70 \, \text{MeV}$ near synchronous altitude. However, pitch-angle diffusion is not known to play an important role in establishing the observed flux profile of outer-zone protons, and currently successful models of the earth's proton radiation environment seldom include pitch-angle diffusion for protons having $E \gtrsim 1 \, \text{MeV}$.

Radial Diffusion. The processes that are known to produce radial diffusion of geomagnetically trapped particles generally involve disturbances

---

[57]However, formation of the slot could also be related to the diminishing importance of radial diffusion at $L \lesssim 3$ and/or to the presence of an internal (CRAND?) source [20] at $L \lesssim 2$. The flux profile of outer-zone electrons (terminating at the slot) can be understood in terms of a balance between pitch-angle diffusion and radial diffusion [70] from a source at $L \gtrsim 6$.

of magnetospheric extent. Sudden magnetic and electrostatic impulses violate only the third invariant of adiabatic motion for radiation-belt particles. The radial-diffusion coefficients D_{LL} produced by these disturbances (whose power spectra vary approximately as ω^{-2}) are found to be strong functions of L. Magnetic sudden impulses typically yield for D_{LL} a magnitude that is independent of particle species and energy, but which is much larger for particles mirroring near the equator than for those mirroring at high magnetic latitudes. Electrostatic sudden impulses produce a D_{LL} whose magnitude varies as the inverse square of the particle drift frequency $\Omega_3/2\pi$. Thus, the value of D_{LL} resulting from electrostatic impulses depends upon particle species and energy, but the dependence of this D_{LL} on equatorial pitch angle is quite weak.

Several empirical estimates for D_{LL} at constant M and $J=0$ are collected in Fig. 83 [122]. The numbered curves correspond to the reference list that follows this chapter. These estimates have been obtained by investigators using a variety of analytical methods, as indicated in the caption. The results reported to apply beyond $L \sim 2$ vary by an order of magnitude up and down from a "compromise" D_{LL} of the form $D_{LL} \approx 1.3 \times 10^{-9} \, L^{10}$ day^{-1}. Thus, there is considerable room for disagreement concerning the "best" numerical value[58] for D_{LL}.

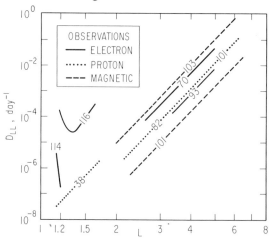

Fig. 83. Compilation [122] of radial diffusion coefficients obtained by various empirical methods, assuming constant M and J: counting of magnetic impulses [101, 103]; spatial quadrature [114, 116]; variational technique [93]; temporal integration [70, 82]; spatial integration [38, 101].

---

[58]In working with algebraic expressions for D_{LL}, it is convenient to remember certain approximate numerical relationships, *e.g.*, $2^{10} = 1024 \sim 10^3$, $4^{10} \sim 10^6$, $8^{10} \sim 10^9$; $3^{10} \sim 6 \times 10^4$, $6^{10} \sim 6 \times 10^7$; $5^{10} \sim 10^7$; $7^{10} \sim 3 \times 10^8$; $\pi^2 \sim 10$, $\pi^3 \approx 31$; 1 year $\sim \pi \times 10^7$ sec; $e^3 \approx 20$, $e^{0.7} \approx 2$, $e^{2.3} \approx 10$.

It is important to realize, when considering the spread of results in Fig. 83, that these values of D_{LL} were obtained for particles having a considerable range of M values. Some of the spread could well result from a genuine variation of D_{LL} with M and/or particle species among the various observations, e.g., from the contribution of electrostatic impulses to radial diffusion. Moreover, the magnitude of D_{LL} surely varies with geomagnetic activity, and hence with time. Finally, discrepancies among several of the reported values of D_{LL} may originate from an incompatibility among the assumptions that underlie the various methods of empirical analysis (in particular [70], [93], and [116]).

The inverse variation of D_{LL} with L for electrons at low L values (see Fig. 83) is apparently a real effect; i.e., not an artifact of the analytical procedure. The enhanced radial diffusion at $L \lesssim 1.2$ may possibly result from pitch-angle diffusion in the presence of drift-shell splitting (see Section III.7). No similar enhancement is evident in D_{LL} for protons [38], which do not suffer significant pitch-angle diffusion from atmospheric collisions. If pitch-angle diffusion is indeed responsible for enhanced electron radial diffusion below $L \approx 1.2$, then the hypothesis of constant M and J in the analysis should be replaced by a hypothesis of constant energy; this would approximately double the required magnitude of D_{LL} in this region. Elsewhere in the radiation belts, the magnitudes of D_{LL} that could result from constant-energy processes are typically found to be too small (by factors ~ 4—100) to account for the particle observations. Moreover, the direction of stochastic flow in L at constant energy is often opposite to the flow direction at constant M and J.

Quo Vadimus? Radial diffusion and pitch-angle diffusion are essential processes in the dynamics of geomagnetically trapped radiation. The determination of numerical values for the various diffusion coefficients will surely continue to be a subject of major scientific interest, since it is important to be able to understand and, ultimately, to predict the evolution of the radiation belts.

Progressively sophisticated methods of data analysis are being brought to bear on the problem of extracting diffusion coefficients from particle observations. One of the next major steps will probably involve the fitting of observational data within the framework of a multi-dimensional Fokker-Planck equation through the use of least-squares techniques. In order for such procedures to be utilized most profitably, the particle distribution functions must be specified with as much detail as is technologically feasible. This objective requires extensive spatial and spectral coverage of the radiation belts by coherently instrumented satellites. In many analyses of this type it will be necessary to augment the basic diffusion equation with terms representing non-diffusive phenomena, e.g., distributed particle sources, particle acceleration *in situ*, particle "injection" associated with magnetic storms, and inelastic collisions with the atmosphere.

Much more effort will be devoted in the future to spectral analysis of the electromagnetic and electrostatic fields present in the radiation belts. The spectra that result from such analyses can be utilized directly to calculate the magnitudes expected of the various diffusion coefficients. Much more effort will probably be devoted as well to relating these *in situ* spectral measurements with similar measurements made at ground stations. Eventually it may become possible to utilize the extensive geographical coverage of ground measurements to infer the condition of the magnetospheric environment. This would make it possible to monitor disturbances from the ground while observing the response of trapped particles in space, with the ultimate goal of gaining predictive insight into the consequences of magnetospheric dynamical processes.

References

1. Van Allen, J. A., Ludwig, G. H., Ray, E. C., McIlwain, C. E.: Observation of high intensity radiation by satellites 1958 alpha and gamma. Jet Propulsion **28**, 588 (1958).
2. Vernov, S, N., Chudakov, A. E., Gorchakov, E. V., Logachev, J. L., Vakulov, P. V.: Study of the cosmic-ray soft component by the third Soviet earth satellite. Planet. Space Sci. **1**, 86 (1959).
3. Northrop, T. G.: The Adiabatic Motion of Charged Particles. New York: Interscience Publishers 1963.
4. Jacobs, J. A.: Geomagnetic Micropulsations. Berlin-Heidelberg-New York: Springer 1970.
5. Roederer, J. G.: Dynamics of Geomagnetically Trapped Radiation. Berlin-Heidelberg-New York: Springer 1970.
6. Omholt, A.: The Optical Aurora. Berlin-Heidelberg-New York: Springer 1971.
7. Goldstein, H.: Classical Mechanics. Reading: Addison-Wesley 1950.
8. Vette, J. I., Lucero, A. B., Wright, J. A.: Inner and Outer Zone Electrons, p. 20. NASA SP-3024 (Washington 1966).
9. Hundhausen, A. J.: Composition and dynamics of the solar wind plasma. Rev. Geophys. Space Phys. **8**, 729 (1970).
10. Spreiter, J. R., Alksne, A. Y., Summers, A. L.: External aerodynamics of the magnetosphere. In: Physics of the Magnetosphere, p. 301. (Ed. Carovillano, R. L., McClay, J. F., Radoski, H. R.) Dordrecht, Holland: D. Reidel 1968.
11. Axford, W. I.: Magnetospheric convection. Rev. Geophys. **7**, 421 (1969).
12. Eviatar, A., Wolf, R. A.: Transfer processes in the magnetopause. J. Geophys. Res. **73**, 5561 (1968); Willis, D. M.: Structure of the magnetopause. Rev. Geophys. Space Phys. **9**, 953 (1971).
13. Chappell, C. R., Harris, K. K., Sharp, G. W.: A study of the influence of magnetic activity on the location of the plasmapause as measured by OGO 5. J. Geophys. Res. **75**, 50 (1970).
14. Brice, N. M.: Bulk motion of the magnetosphere. J. Geophys. Res. **72**, 5193 (1967).
15. Frank, L. A.: On the extraterrestrial ring current during geomagnetic storms. J. Geophys. Res. **72**, 3753 (1967).
16. Serbu, G. P., Maier, E. J. R.: Observations from OGO 5 of the thermal ion density and temperature within the magnetosphere. J. Geophys. Res. **75**, 6102 (1970).
17. Smart, D. F., Shea, M. A., Gall, R.: The daily variation of trajectory-derived high-latitude cutoff rigidities in a model magnetosphere. J. Geophys. Res. **74**, 4731 (1969).
18. Lehnert, B.: Dynamics of Charged Particles. Amsterdam: North-Holland 1964.

19. Stern, D. P.: Euler potentials and geomagnetic drift shells. J. Geophys. Res.
    73, 4373 (1968); Stern, D. P.: Euler potentials. Amer. J. Phys. **38**, 494 (1970);
    Stern, D. P.: Shell splitting due to electric fields. J. Geophys. Res. **76**, 7787
    (1971).
20. Lenchek, A. M., Singer, S. F., Wentworth, R. C.: Geomagnetically trapped
    electrons from cosmic ray albedo neutrons. J. Geophys. Res. **66**, 4027 (1961).
21. McIlwain, C. E.: Magnetic coordinates. Space Sci. Rev. **5**, 585 (1966).
22. Hilton, H. H.: L parameter, A new approximation. J. Geophys. Res. **76**,
    6952 (1971).
23. Cole, K. D.: Magnetic storms and associated phenomena. Space Sci. Rev.
    5, 699 (1966).
24. Hoffman, R. A., Bracken, P. A.: Higher-order ring currents and particle
    energy storage in the magnetosphere. J. Geophys. Res. **72**, 6039 (1967).
25. Chandrasekhar, S.: Plasma Physics. Chicago: Univ. of Chicago Press 1960.
26. Söraas, F., Davis, L. R.: Temporal variations of the 100 keV to 1700 keV
    trapped protons observed on satellite Explorer 26 during first half of 1965.
    GSFC X-612-68-328 (Greenbelt 1968).
27. Hones, E. W.: Motion of charged particles trapped in the earth's magneto-
    sphere. J. Geophys. Res. **68**, 1209 (1963).
28. Mead, G. D.: Deformation of the geomagnetic field by the solar wind.
    J. Geophys. Res. **69**, 1181 (1964).
29. Schulz, M., Eviatar, A.: Diffusion of equatorial particles in the outer radiation
    zone. J. Geophys. Res. **74**, 2182 (1969).
30. Williams, D. J., Mead, G. D.: Night-side magnetospheric configuration as
    obtained from trapped electrons at 1100 kilometers. J. Geophys. Res. **70**,
    3017 (1965).
31. Roederer, J. G.: Quantitative models of the magnetosphere. Rev. Geophys.
    7, 77 (1969).
32. Birmingham, T. J., Jones, F. C.: Identification of moving magnetic field
    lines. J. Geophys. Res. **73**, 5505 (1968).
33. Taylor, H. E., Hones, E. W.: Adiabatic motion of auroral particles in a
    model of the electric and magnetic fields surrounding the earth. J. Geophys.
    Res. **70**, 3605 (1965).
34. Schulz, M.: Compressible corotation of a model magnetosphere. J. Geophys.
    Res. **75**, 6329 (1970).
35. Vasyliunas, V.: A crude estimate of the relation between the solar wind
    speed and the magnetospheric electric field. J. Geophys. Res. **73**, 2529 (1968).
36. Hamlin, D. A., Karplus, R., Vik, R. C., Watson, K. M.: Mirror and azimuthal
    drift frequencies for geomagnetically trapped particles. J. Geophys. Res. **66**,
    1 (1961); Schulz, M.: Approximate second invariant for a dipole field. J.
    Geophys. Res. **76**, 3144 (1971).
37. Evans, R. D.: The Atomic Nucleus, p. 637. New York: McGraw-Hill 1955.
38. Farley, T. A., Tomassian, A. D., Walt, M.: Source of high-energy protons
    in the Van Allen radiation belt. Phys. Rev. Letters **25**, 47 (1970); Farley,
    T. A., Walt, M.: Source and loss processes of protons of the inner radiation
    belt. J. Geophys. Res. **76**, 8223 (1971).
39. Schulz, M., Paulikas, G. A.: Secular magnetic variation and the inner proton
    belt. J. Geophys. Res. **77**, 744 (1972); Heckman, H. H., Lindstrom, P. J.:
    Response of trapped particles to a collapsing dipole moment. J. Geophys.
    Res. **77**, 740 (1972).
40. Cornwall, J. M.: Transport and loss processes for magnetospheric helium.
    J. Geophys. Res. **76**, 264 (1971); Cornwall, J. M.: Radial diffusion of ionized

helium and protons: A probe for magnetospheric dynamics. J. Geophys. Res. **77**, 1756 (1972); Claflin, E. S.: Charge-exchange cross sections for hydrogen and helium ions incident on atomic hydrogen: 1 to 1000 keV. SAMSO TR-70-258 (Los Angeles 1970).

41. Johnson, F. S.: Atmospheric structure. Astronautics **7**, 54 (1962); Cornwall, J. M., Sims, A. R., White, R. S.: Atmospheric density experienced by radiation belt protons. J. Geophys. Res. **70**, 3099 (1965).

42. Walt, M.: Loss rates of trapped electrons by atmospheric collisions. In: Radiation Trapped in the Earth's Magnetic Field, p. 337. (Ed. McCormac, B. M.) Dordrecht, Holland: D. Reidel 1966.

43. Roberts, C. S.: Pitch-angle diffusion of electrons in the magnetosphere. Rev. Geophys. **7**, 305 (1969).

44. Hasegawa, A.: Heating of the magnetospheric plasma by electromagnetic waves generated in the magnetosheath. J. Geophys. Res. **74**, 1763 (1969).

45. Helliwell, R. A.: Whistlers and Related Ionospheric Phenomena. Stanford: Stanford Univ. Press 1965; Helliwell, R. A.: Low-frequency waves in the magnetosphere. Rev. Geophys. **7**, 281 (1969).

46. Barfield, J. N., Lanzerotti, L. J., Maclennan, C. G., Paulikas, G. A., Schulz, M.: Quiettime observation of a coherent compressional Pc-4 micropulsation at synchronous altitude. J. Geophys. Res. **76**, 5252 (1971).

47. Russell, C. T., Holzer, R. E.: AC magnetic fields. In: Particles and Fields in the Magnetosphere, p. 195. (Ed. McCormac, B. M.) Dordrecht, Holland: . D. Reidel 1970; Campbell, W. H.: Geomagnetic pulsations. In: Physics of Geomagnetic Phenomena, p. 821. (Ed. Matsushita, S., Campbell, W. H.) New York: Academic Press 1967.

48. Roberts, C. S., Schulz, M.: Bounce resonant scattering of particles trapped in the earth's magnetic field. J. Geophys. Res. **73**, 7361 (1968).

49. Stix, T. H.: The Theory of Plasma Waves. New York: McGraw-Hill 1962.

50. Rowlands, J., Shapiro, V. D., Shevchenko, V. I.: Quasilinear theory of plasma cyclotron instability. Soviet Physics JETP **23**, 651 (1966).

51. Bailey, D. K.: Some quantitative aspects of electron precipitation in and near the auroral zone. Rev. Geophys. **6**, 289 (1968); Thorne, R. M., Kennel, C. F.: Relativistic electron precipitation during magnetic storm main phase. J. Geophys. Res. **76**, 4446 (1971); Vampola, A. L.: Electron pitch angle scattering in the outer zone during magnetically disturbed times. J. Geophys. Res. **76**, 4685 (1971).

52. Kennel, C. F., Petschek, H. E.: Limit on stably trapped particle fluxes. J. Geophys. Res. **71**, 1 (1966); Cornwall, J. M.: Micropulsations and the outer radiation zone. J. Geophys. Res. **71**, 2185 (1966).

53. Kennel, C. F.: Consequences of a magnetospheric plasma. Rev. Geophys. **7**, 379 (1969); Kennel, C. F., Petschek, H. E.: Van Allen belt plasma physics. In: Proceedings of the Second Orsay Summer Institute on Plasma Physics, p. 95. (Ed. Kalman, G., Feix, M.) London: Gordon and Breach 1969.

54. Cornwall, J. M., Coroniti, F. V., Thorne, R. M.: Turbulent loss of ring current protons. J. Geophys. Res. **75**, 4699 (1970); Eather, R. H., Carovillano, R. L.: The ring current as the source region for proton auroras. Cosmic Electrodyn. **2**, 105 (1971).

55. Brewer, H. R., Schulz, M., Eviatar, A.: Origin of drift-periodic echoes in outer-zone electron flux. J. Geophys. Res. **74**, 159 (1969).

56. Fälthammar, C.-G.: Radial diffusion by violation of the third adiabatic invariant. In: Earth's Particles and Fields, p. 157. (Ed. McCormac, B. M.) New York: Reinhold 1968.

57. Kosik, J. Cl.: Diffusion radiale des particules chargées par violation du troisième invariant adiabatique. Ann. Géophys. **27**, 27 (1971).
58. Kellogg, P. J.: Van Allen radiation of solar origin. Nature **183**, 1295 (1959); Parker, E. N.: Geomagnetic fluctuations and the form of the outer zone of the Van Allen radiation belt. J. Geophys. Res. **65**, 3117 (1960).
59. Krimigis, S. M.: Alpha particles trapped in the earth's magnetic field. In: Particles and Fields in the Magnetosphere, p. 364. (Ed. McCormac, B. M.) Dordrecht, Holland: D. Reidel 1970.
60. Hasegawa, A.: Plasma instabilities in the magnetosphere. Rev. Geophys. Space Phys. **9**, 703 (1971).
61. Rose, D. J., Clark, M.: Plasmas and Controlled Fusion, p. 145. Cambridge: M. I. T. Press 1961.
62. Bohm, D.: Qualitative description of the arc plasma in a magnetic field. In: The Characteristics of Electrical Discharges in Magnetic Fields, p. 1. (Ed. Guthrie, A., Wakerling, R. K.) New York: McGraw-Hill 1949.
63. Cornwall, J. M., Coroniti, F. V., Thorne, R. M.: Unified theory of SAR arc formation at the plasmapause. J. Geophys. Res. **76**, 4428 (1971).
64. Dungey, J. W.: Effects of electromagnetic perturbations on particles trapped in the radiation belts. Space Sci. Rev. **4**, 199 (1965).
65. Fälthammar, C.-G., Walt, M.: Radial motion resulting from pitch-angle scattering of trapped electrons in the distorted geomagnetic field. J. Geophys. Res. **74**, 4184 (1969); Roederer, J. G., Schulz, M.: Effect of shell splitting on radial diffusion in the magnetosphere. J. Geophys. Res. **74**, 4117 (1969).
66. Schulz, M.: Drift-shell splitting at arbitrary pitch angle. J. Geophys. Res. **77**, 624 (1972).
67. Roederer, J. G.: Geomagnetic field distortions and their effects on radiation belt particles. Rev. Geophys. Space Phys. **10**, 599 (1972); Roederer, J. G., Hilton, H. H., Schulz, M.: Drift-shell splitting by internal geomagnetic multipoles. J. Geophys. Res. **78**, 133 (1973).
68. Haerendel, G.: Diffusion theory of trapped particles and the observed proton distribution. In: Earth's Particles and Fields, p. 171. (Ed. McCormac, B. M.) New York: Reinhold 1968.
69. Theodoridis, G. C.: Bimodal diffusion in the earth's magnetosphere: 1. An acceleration mechanism for trapped particles. Ann. Géophys. **24**, 944 (1968); Theodoridis, G. C., Paolini, F. R., Frankenthal, S.: Bimodal diffusion in the earth's magnetosphere: 2. On the electron belts. Ann. Géophys. **24**, 1015 (1968).
70. Newkirk, L. L., Walt, M.: Radial diffusion coefficient for electrons at $1.76 < L < 5$. J. Geophys. Res. **73**, 7231 (1968).
71. Walt, M.: Radial diffusion of trapped particles. In: Particles and Fields in the Magnetosphere, p. 410. (Ed. McCormac, B. M.) Dordrecht, Holland: D. Reidel 1970.
72. Lingenfelter, R. E.: The cosmic-ray neutron leakage flux. J. Geophys. Res. **68**, 5633 (1963).
73. Dragt, A. J., Austin, M. M., White, R. S.: Cosmic ray and solar proton albedo neutron decay injection. J. Geophys. Res. **71**, 1293 (1966); Dragt, A. J.: Solar cycle modulation of the radiation belt proton flux. J. Geophys. Res. **76**, 2313 (1971).
74. Walt, M.: The effects of atmospheric collisions on geomagnetically trapped electrons. J. Geophys. Res. **69**, 3947 (1964); Walt, M., Newkirk, L. L.: Addition to investigation of the decay of the Starfish radiation belt. J. Geophys. Res. **71**, 3265 (1966).

75. Brown, W. L.: Observations of the transient behavior of electrons in the artificial radiation belts. In: Radiation Trapped in the Earth's Magnetic Field, p. 612. (Ed. McCormac, B. M.) Dordrecht, Holland: D. Reidel 1966.
76. Imhof, W. L., Smith, R. V.: Observation of nearly mono-energetic high-energy electrons in the inner radiation belt. Phys. Rev. Letters 14, 885 (1965).
77. Imhof, W. L.: Electron precipitation in the radiation belts. J. Geophys. Res. 73, 4167 (1968).
78. Filz, R. C., Holeman, E.: Time and altitude dependence of 55-MeV trapped protons, August 1961 to June 1964. J. Geophys. Res. 70, 5807 (1965).
79. Harris, I., Priester, W.: Time dependent structure of the upper atmosphere. GSFC X-640-62-69 (Greenbelt 1962); Harris, I., Priester, W.: Theoretical models for the solar-cycle variation of the upper atmosphere. GSFC X-640-62-70 (Greenbelt 1962).
80. Williams, D. J., Arens, J. F., Lanzerotti, L. J.: Observations of trapped electrons at low and high altitudes. J. Geophys. Res. 73, 5673 (1968).
81. Paulikas, G. A., Blake, J. B.: Effects of sudden commencements on solar protons at the synchronous orbit. J. Geophys. Res. 75, 734 (1970).
82. Söraas, F.: Comparison of post-storm non-adiabatic recovery of trapped protons with radial diffusion. GSFC X-612-69-241 (Greenbelt 1969).
83. Paulikas, G. A., Blake, J. B., Palmer, J. A.: Energetic electrons at the synchronous altitude: A compilation of data, p. 10. SAMSO TR-69-413 (Los Angeles 1969).
84. Davis, L. R., Williamson, J. M.: Low-energy trapped protons. Space Res. 3, 365 (1963).
85. Mihalov, J. D., White, R. S.: Low-energy proton radiation belts. J. Geophys. Res. 71, 2207 (1966).
86. Nakada, M. P., Dungey, J. W., Hess, W. N.: On the origin of outer-belt protons, 1. J. Geophys. Res. 70, 3529 (1965).
87. Fritz, T. A., Williams, D. J.: Initial observations of geomagnetically trapped alpha particles at the equator. J. Geophys. Res. 78, 4719 (1973).
88. Lanzerotti, L. J., Robbins, M. F.: Solar flare alpha to proton ratio changes following interplanetary disturbances. Solar Phys. 10, 212 (1969).
89. Axford, W. I.: On the origin of radiation belt and auroral primary ions. In: Particles and Fields in the Magnetosphere, p. 46. (Ed. McCormac, B. M.) Dordrecht, Holland: D. Reidel 1970.
90. Paolini, F. R., Theodoridis, G. C., Frankenthal, S., Katz, L.: Radial diffusion processes of relativistic outer-belt electrons. Ann. Géophys. 24, 129 (1968).
91. Pfitzer, K. A., Lezniak, T. W., Winckler, J. R.: Experimental verification of drift-shell splitting in the distorted magnetosphere. J. Geophys. Res. 74, 4687 (1969).
92. McDiarmid, I. B., Burrows, J. R.: Dependence of the position of the outer radiation zone intensity maxima on electron energy and magnetic activity. Canad. J. Phys. 45, 2873 (1967).
93. Lanzerotti, L. J., Maclennan, C. G., Schulz, M.: Radial diffusion of outer-zone electrons: An empirical approach to third-invariant violation. J. Geophys. Res. 75, 5351 (1970).
94. Vernov, S. N., Kuznetsov, S. N., Logachev, Yu. I., Lopatina, G. B., Sosnovets, E. N., Stolpovskiy, V. G.: Radial diffusion of electrons of energy greater than 100 keV in the outer radiation belt. Geomag. Aeron. 8, 323 (1968).
95. Pfitzer, K. A., Winckler, J. R.: Experimental observation of a large addition to the electron inner radiation belt following a solar flare event. J. Geophys. Res. 73, 5792 (1968).

96. Rosen, A., Sanders, N. L.: Loss and replenishment of electrons in the inner radiation zone during 1965—1967. J. Geophys. Res. **76**, 110 (1971).

97. Craven, J. D.: Temporal variations of electron intensities at low altitudes in the outer radiation zone as observed with satellite Injun 3. J. Geophys. Res. **71**, 5643 (1966).

98. Taylor, W. W. L., Gurnett, D. A.: The morphology of VLF emissions observed with the Injun 3 satellite. J. Geophys. Res. **73**, 5616 (1968).

99. Vampola, A. L., Koons, H. C., McPherson, D. A.: Outer-zone electron precipitation. J. Geophys. Res. **76**, 7609 (1971).

100. Lanzerotti, L. J., Tartaglia, N. A.: Propagation of a magnetospheric compressional wave to the ground. J. Geophys. Res. **77**, 1934 (1972).

101. Nakada, M. P., Mead, G. D.: Diffusion of protons in the outer radiation belt. J. Geophys. Res. **70**, 4777 (1965).

102. Tverskoy, B. A.: Dynamics of the radiation belts of the earth. Geomag. Aeron. **3**, 351 (1964).

103. Tverskoy, B. A.: Transport and acceleration of charged particles in the earth's magnetosphere. Geomag. Aeron. **5**, 517 (1965).

104. Nishida, A.: Satellite information on the origin of geomagnetic variations. Ann. Géophys. **26**, 401 (1970).

105. Romañá, A.: Geomagnetic and solar data: International data on magnetic disturbances. J. Geophys. Res. **64**, 1349 (1959).

106. Davidson, M. J.: Average diurnal characteristics of geomagnetic power spectra in the period range 4.5 to 1000 seconds. J. Geophys. Res. **69**, 5116 (1964).

107. Nishida, A., Cahill, L. J.: Sudden impulses in the magnetosphere observed by Explorer 12. J. Geophys. Res. **69**, 2243 (1964); Patel, V. L., Coleman, P. J.: Sudden impulses in the magnetosphere observed at synchronous orbit. J. Geophys. Res. **75**, 7255 (1970).

108. Lanzerotti, L. J., Roberts, C. S., Brown, W. L.: Temporal variations in the electron flux at synchronous altitudes. J. Geophys. Res. **72**, 5893 (1967).

109. Brown, W. L.: Energetic outer-belt electrons at synchronous altitude. In: Earth's Particles and Fields, p. 33. (Ed. McCormac, B. M.) New York: Reinhold 1968.

110. Lanzerotti, L. J., Maclennan, C. G., Robbins, M. F.: Proton drift echoes in the magnetosphere. J. Geophys. Res. **76**, 259 (1971).

111. Frank, L. A.: Inward radial diffusion of electrons greater than 1.6 million electron volts in the outer radiation zone. J. Geophys. Res. **70**, 3533 (1965).

112. McDiarmid, I. B., Burrows, J. R.: Temporal variations of outer radiation zone electron intensities at 1000 km. Canad. J. Phys. **44**, 1361 (1966).

113. Tomassian, A. D., Farley, T. A., Vampola, A. L.: Inner-zone energetic-electron repopulation by radial diffusion. J. Geophys. Res. **77**, 3441 (1972).

114. Newkirk, L. L., Walt, M.: Radial diffusion coefficient for electrons at low *L* values. J. Geophys. Res. **73**, 1013 (1968).

115. Farley, T. A.: Radial diffusion of electrons at low *L* values. J. Geophys. Res. **74**, 377 (1969).

116. Farley, T. A.: Radial diffusion of starfish electrons. J. Geophys. Res. **74**, 3591 (1969); Chapman, M. C., Farley, T. A.: Absolute electron fluxes and energies in the inner radiation zone in 1965. J. Geophys. Res. **73**, 6825 (1968).

117. Kavanagh, L. D.: An empirical evaluation of radial diffusion coefficients for electrons of 50—100 keV from $L=4$ to $L=7$. J. Geophys. Res. **73**, 2959 (1968).

118. Carter, R. E., Reines, F., Wagner, J. J., Wyman, M. E.: Free antineutrino absorption cross section. II. Expected cross section from measurements of fission fragment electron spectrum. Phys. Rev. **113**, 280 (1959).
119. Walt, M., Newkirk, L. L.: Comments on 'Radial diffusion of outer-zone electrons'. J. Geophys. Res. **76**, 5368 (1971); Lanzerotti, L. J., Maclennan, C. G. Schulz, M.: Reply. J. Geophys. Res. **76**, 5371 (1971).
120. Preszler, A. M., Simnett, G. M., White, R. S.: Earth albedo neutrons from 10 to 100 MeV. Phys. Rev. Letters **28**, 982 (1972).
121. Lyons, L. R., Thorne, R. M., Kennel, C. F.: Pitch-angle diffusion of radiation-belt electrons within the plasmasphere. J. Geophys. Res. **77**, 3455 (1972).
122. Walt, M.: Radial diffusion of trapped particles and some of its consequences. Rev. Geophys. Space Phys. **9**, 11 (1971); Walt, M.: The radial diffusion of trapped particles induced by fluctuating magnetospheric fields. Space Sci. Rev. **12**, 446 (1971).
123. Imhof, W. L., Reagan, J. B., Smith, R. V.: Long-term study of electrons trapped on low L shells. J. Geophys. Res. **72**, 2371 (1971).

Frequently used Symbols

A, A_j	atomic weight (number of nucleons in ion)
$\mathbf{A}$	magnetic vector potential
a	earth radius (6371.2 km)
$\mathbf{B}$	magnetic field
B	magnitude of $\mathbf{B}$
$\hat{\mathbf{B}}$	unit vector $\mathbf{B}/B$
B_0, B_1, B_2	coefficients of Mead field
B_e, B_m	equatorial field, mirror field
B_r, B_θ, B_φ	components of $\mathbf{B}$ (spherical coordinates)
B_t	tail-field intensity
$\mathbf{b}$	magnetic-field perturbation
b	stand-off distance; magnitude of $\mathbf{b}$
$b_\perp, b_\parallel$	components of $\mathbf{b}$ (relative to $\hat{\mathbf{B}}$)
b_x, b_y	components of $b_\perp$
$\mathscr{B}_\parallel, \mathscr{B}_\perp, \mathscr{B}_z$	magnetic spectral densities
c	speed of light (3×10^5 km/sec)
c_A	Alfvén speed
$D, D(y)$	dipole drift function, $= (1/2)\,T(y) - (1/12)\,Y(y)$
$\mathbb{D}, \underset{\sim}{\mathbb{D}}$	diffusion tensors
D_{xx}	pitch-angle (-cosine) diffusion coefficient
D_{LL}	radial diffusion coefficient
D_{LL}^*	Bohm diffusion coefficient
D_i'	Fokker-Planck coefficient
D_{st}	ring-current index
$\mathbf{E}$	electric field
E	kinetic energy
E_c	convection electric field (equatorial magnitude)
E_0	e-folding energy
E_r, E_θ, E_φ	components of $\mathbf{E}$ (spherical coordinates)
$\mathbf{e}$	electric-field perturbation
$\mathscr{E}_c, \mathscr{E}_m$	electrostatic spectral densities
$\mathscr{E}_\perp, \mathscr{E}_\parallel$	electrostatic spectral densities
$\mathbf{F}$	force; guiding-center force
F_j	unit-normalized distribution function

f, f_j	phase-space distribution function
$\bar{f}, \bar{f}_j$	phase average of f (j identifies particle species)
$\tilde{f}, \tilde{f}_j$	perturbation of f, i. e., $\tilde{f} = f - \bar{f}$
$f_\|, f_\|(s,t)$	perturbing force (parallel to $\mathbf{B}$)
f_n	Fourier amplitude of $f_\|(s,t)$
$\mathscr{F}_\|$	spectral density of $f_\|(s,t)$
$G(J_i; Q_j)$	Jacobian, $= \det(\partial J_i / \partial Q_j)$
$G_1(y), G_2(y)$	shell-tracing functions
$g_n(x)$	pitch-angle eigenfunction ($n = 0, 1, 2, \ldots$)
H	Hamiltonian function
$\hbar$	Planck's constant ($\div 2\pi$)
I	$J/2p$, geometrical "invariant"; electrical current
I_i	mean excitation energy
$I_\alpha, I_\perp$	integral unidirectional flux ($I_\perp \equiv I_{\pi/2}$)
$I_{4\pi}$	integral omnidirectional flux
$I_{4\pi}^*, I^*$	maximum integral omnidirectional flux
Im	imaginary part
J	second adiabatic invariant
J_i	fundamental action integral ($i = 1, 2, 3$)
$J_\alpha, J_\perp$	differential unidirectional flux ($J_\perp = J_{\pi/2}$)
$\bar{J}_\alpha, \bar{J}_\perp$	phase averages of J_α and $J_\perp$
$J_{4\pi}$	differential omnidirectional flux
J_0, J_l	Bessel function (order zero, order l)
$\mathbf{J}$	electrical current density
K	derived invariant, $= J/(8 m_0 M)^{1/2}$
K_p	index of geomagnetic activity
$\mathbf{k}$	wave-propagation vector
L	invariant drift-shell parameter, $= 2\pi B_0 a^2 |\Phi|^{-1}$
L_d	field-line label
L_m	McIlwain shell parameter
M	first adiabatic invariant
m	relativistic mass; harmonic number
m_0, m_j	rest mass (j identifies particle species)
N_e, N_p	electron and proton densities (particles/cm^3)
$\bar{N}_e, \bar{N}_i$	atmospheric densities (drift averaged)
n	refractive index, $= c k_\| / \omega$; harmonic number
P_l^m	associated Legendre function (Schmidt-normalized)
$P_\perp, P_\|$	plasma pressures (relative to $\hat{\mathbf{B}}$)
$\mathbf{p}$	particle momentum
$Q(y)$	shell-tracing function
Q_i	new coordinate (generalized)
$\mathbf{q}, q_i$	canonical position ($i = 1, 2, 3$)
q	particle charge; charge-exchange (subscript)

Re	real part
r	radial coordinate
S	arc length of field line; source for $\partial \bar{f}/\partial t$
$T, T(y)$	dipole bounce function
$\tilde{T}(y; L_d, \varphi)$	bounce function in distorted field
t	time
$\mathbf{u}$	solar-wind velocity
V_e, V_m	electric and magnetic scalar potentials
$\mathbf{v}_d$	guiding-center drift velocity
v_g, v_p	group velocity ($d\omega/dk_{\parallel}$), phase velocity ($\omega/k_{\parallel}$)
$\mathbf{v}$	particle velocity
$\mathscr{V}, \mathscr{V}_m$	spectral density of electrostatic potential
W	energy (kinetic plus potential), $= E + q V_e(\mathbf{r})$
w	energy-like variable, $p^2/2m_0$
X	field-line coordinate, $X^2 \equiv 1 - (B_e/B)$
x	cosine of equatorial pitch angle
x_b, x_c	cosine of (bounce-, drift-) loss-cone half-angle
$Y, Y(y)$	dipole bounce function
$\tilde{Y}(y; L_d, \varphi)$	bounce function in distorted field
y	sine of equatorial pitch angle
Z, Z_i	ionic charge number, nuclear charge number
α	local pitch angle; Euler potential
β	Euler potential
$\beta_{\perp}, \beta_{\parallel}$	plasma indices ($8\pi P_{\perp}/B^2$ and $8\pi P_{\parallel}/B^2$)
$\Gamma(s+1)$	gamma function, $s!$
γ, γ_j	ratio of relativistic mass to rest mass
γ	unit of magnetic intensity, $= 10^{-5}$ gauss
δ_{ij}	Kronecker symbol ($= 1$ if $i=j$; $= 0$ otherwise)
ε	adiabaticity index, $= \langle v/\Omega_1 S \rangle$
$\varepsilon_1, \varepsilon_2$	field-expansion indices, $\varepsilon_l \equiv (B_l/B_0)(L_d a/b)^{l+2}$
ζ	M/y^2, approximate invariant
θ	colatitude; angle between $\hat{\mathbf{k}}$ and $\hat{\mathbf{B}}$
θ_m	colatitude of mirror point
Λ	"invariant" magnetic latitude ($L = \sec^2 \Lambda$)
λ_D	Debye length
$\mu_{\parallel}, \mu_{\perp}, \mu_{\rfloor}$	electrical mobilities (Ohm, Pedersen, Hall)
π, π_i	canonical momentum ($i = 1, 2, 3$)
ρ_s	solar-wind mass density
τ	interaction time; pitch-angle-diffusion lifetime
Φ	third adiabatic invariant
φ, φ_a	magnetic longitude, longitude of "anomaly"
φ_i	phase conjugate to $J_i/2\pi$ ($i = 1, 2, 3$)

ψ_n	Fourier phase of wave or field
ψ_s	angle of deflection (solar wind)
$\Omega_i/2\pi$	frequency of gyration, bounce, or drift ($i = 1, 2, 3$)
Ω_j	nonrelativistic gyrofrequency ($\times 2\pi$), $-q_j B/m_j c$
$\mathbf{\Omega}_0$	angular velocity of earth's rotation
ω	frequency ($\times 2\pi$)
ω_j	ion or electron plasma frequency ($\times 2\pi$)
$\omega_n/2\pi$	Fourier frequency of wave or field ($n = 1, 2, 3, ...$)

Subject Index

Physics and Chemistry in Space

Edited by J. G. Roederer,
Denver, Colo.

Editorial Board:
H. Elsässer, Heidelberg;
G. Elwert, Tübingen;
L. G. Jacchia,
Cambridge, Mass.;
J. A. Jacobs, Edmonton,
Alta.;
N. F. Ness, Greenbelt;
Md.;
W. Riedler, Graz

A series of monographs
written and published
to serve the student,
the teacher and the
researcher with a clear
and concise presentation
of up-to-date topics of
space exploration.

Vol. 1: Jacobs Geomagnetic Micropulsations

By J. A. Jacobs, Killam
Memorial, Professor of
Science, The University of
Alberta, Edmonton, Canada
With 81 figs. VIII, 179 pp. 1970
Cloth DM 36,–; US $ 13.90
ISBN 3-540-04986-X

Vol. 2: Roederer Dynamics of Geomagnetically Trapped Radiation

By J. G. Roederer, Professor
of Physics, University of
Denver, Denver, Colo., USA
With 94 figs. XIV, 166 pp. 1970
Cloth DM 36,–; US $ 13.90
ISBN 3-540-04987-8

Vol. 3: Adler / Trombka Geochemical Exploration of the Moon and Planets

By Dr. I. Adler, Senior Scien-
tist, and Dr. J. I. Trombka,
both: Goddard Space Flight
Center, NASA, Greenbelt,
Md., USA
With 129 figs. X, 243 pp. 1970
Cloth DM 58,–; US $ 22.40
ISBN 3-540-05228-3

Vol. 4: Omholt The Optical Aurora

By A. Omholt, Professor of
Physics, Universitetet i Oslo,
Fysisk Institutt, Blindern,
Oslo, Norway
With 54 figs. XIII, 198 pp. 1971
Cloth DM 58,–; US $ 22.40
ISBN 3-540-05486-3

Vol. 5: Hundhausen Coronal Expansion and Solar Wind

By A. J. Hundhausen,
High Altitude Observatory,
National Center for
Atmospheric Research,
Boulder, Colo. USA
With 101 figs. XII, 238 pp. 1972
Cloth DM 68,–; US $ 26.20
ISBN 3-540-05875-3

Vol. 6: Bauer Physics of Planetary Ionospheres

By S. J. Bauer, Laboratory
for Planetary Atmospheres,
NASA Goddard Space Flight
Center, Greenbelt, Md, USA
With 89 figs. VIII, 230 pp. 1973
Cloth DM 78,–; US $ 30.10
ISBN 3-540-06173-8

Prices are subject to change
without notice.

Springer-Verlag
Berlin
Heidelberg
New York

München   Johannesburg
London   New Delhi
Paris   Rio de Janeiro
Sydney   Tokyo   Wien

Astrophysics

13 figs. IV, 120 pages. 1973
(Springer Tracts in Modern Physics, Vol. 69)
Cloth DM 78,–; US $ 30.10
ISBN 3–540–06376–5
Prices are subject to change without notice

Contents

(1) This introductory account encompasses recent results and will be of interest to physicists of various specialties since the description of the outer layers of neutron stars involves solid-state and nuclear physics and that of their interior elementary-particle physics.

(2) Much has been published recently on "black holes" in the general theory of relativity. New findings are summarized with emphasis on potential applications in astrophysics. Topics treated include paths of stars in proximity to the hole, tidal forces, and gravitational waves.

Springer-Verlag
Berlin
Heidelberg
New York
München   Johannesburg
London   New Delhi
Paris   Rio de Janeiro
Sydney   Tokyo   Wien